# Pragmatic Logic

Pragmatic Logic
William J. Eccles

ISBN: 978-3-031-79769-9   paperback
ISBN: 978-3-031-79770-5   ebook

DOI: 10.1007/978-3-031-79770-5

A Publication in the Springer series
*SYNTHESIS LECTURES DIGITAL CIRCUITS AND SYSTEMS #13*

Lecture
Series Editor: Mitchell Thornton, Southern Methodist University

Library of Congress Cataloging-in-Publication Data

Series ISSN:   1932-3166   print
Series ISSN:   1932-3174   electronic

First Edition
10 9 8 7 6 5 4 3 2 1

# Pragmatic Logic

William J. Eccles
Rose-Hulman Institute of Technology

*SYNTHESIS LECTURES ON DIGITAL CIRCUITS AND SYSTEMS #13*

## ABSTRACT

*Pragmatic Logic* presents the analysis and design of *digital logic* systems. The author begins with a brief study of *binary* and *hexadecimal* number systems and then looks at the basics of *Boolean algebra*. The study of logic circuits is divided into two parts, *combinational logic*, which has no memory, and *sequential logic*, which does. Numerous examples highlight the principles being presented. The text ends with an introduction to *digital logic* design using *Verilog*, a hardware description language.

The chapter on *Verilog* can be studied along with the other chapters in the text. After the reader has completed *combinational logic* in Chapters 4 and 5, sections 9.1 and 9.2 would be appropriate. Similarly, the rest of Chapter 9 could be studied after completing *sequential logic* in Chapters 6 and 7.

This short lecture book will be of use to students at any level of electrical or computer engineering and for practicing engineers or scientists in any field looking for a practical and applied introduction to *digital logic*. The author's "pragmatic" and applied style gives a unique and helpful "non-idealist, practical, opinionate" introduction to digital systems.

## KEYWORDS

Digital logic, number systems, Boolean algebra, combinational logic, sequential logic, Verilog

# Contents

# Foreword

The word *pragmatic* means all of those things that I have included in the subtitle. Some people say that *opinionated* fits me very well. But I think the *practical* aspect of the definition is more in line with what I have written here.

*Pragmatism* is an approach to understanding in philosophy that says you can find the meaning of something by looking at its practical side. It also says that thought is around to guide our actions, and then goes further to argue that truth is to be primarily tested by the practical consequences of a belief or theory.

So *Pragmatic Logic* is very much a practical approach to the study of digital logic. I emphasize practical problems so that from these you will learn some of the underlying truths.

The result is a text that does not spend much time on theorems in Boolean algebra. It does not do a lot of drill work in number systems. Instead, we'll spend our time learning some basic techniques and then just doing design. After all, the end result of an engineering effort is the design of something that will benefit people.

Will you be able to design something useful by the end of the course? Yes, I think so. But don't leave with the idea that you really know this field! That will come later with experience. In other words, you'll know just enough to be dangerous!

One practical matter: all of the examples have been done in LogicWorks®, an educations simulation tool for both PCs and Macs, which is available from Amazon.com.

A lot of the author's education and background goes into a text like this, and I am probably no exception. My introduction to digital logic was as an undergraduate in the fifties. The very first textbook on this topic was published then, *Design of Switching Circuits* by Kiester, Ritchie, and Washburn. Until then, this field had not been "regularized" with a text. That book dealt with the design of relay circuits, since electronic switching was fairly uncommon. The authors were at the Bell Telephone Laboratories, then a cauldron of invention and development—the transistor had come out of BTL only about three years before this textbook.

I've taught this material many times since then, out of lots of different textbooks. But those courses were generally a year long! We did combinational logic in the first semester, sequential in the second. Then computers became more available and we condensed this to one semester.

So with thanks to a number of colleagues for helpful conversations, apologies to the reader for any errors that I've made, and relief that this task is about finished, I hope you enjoy learning about digital systems as much as I have enjoyed writing about them.

W.J.E. Terre Haute, Indiana

January 2007

CHAPTER 1

# Designing Digital Systems: Where are We Going?

How many times in the last couple of hours have you encountered a digital system? Is there one on your wrist? Your computer? The innards of your VCR? The microwave oven? Your stereo? When you think about it, just about everything we deal with today has something related to digital logic inside it.

What does *digital* mean? Very simply, it means that the device or the system works with digits, numbers, discrete quantities. These numbers are usually, at least in the interior of the device, binary numbers—0s and 1s. We usually see the results as "people" numbers—decimal, most likely—and alphanumeric characters, including letters and digits and punctuation.

So what's the opposite of *digital? Analog*. This means that the range of values being processed is continuous. A good example is in your car. The speedometer is an analog device because the pointer can point absolutely anywhere on the scale (unless you happen to have a fully digital display as some cars do today). The odometer, on the other hand, is digital because it can display only numbers in a fixed range. Miles to the nearest integer, or perhaps to the nearest tenth. It can't display, say, 2783.9017 miles but instead displays 2783.9 or even 2783.

And where are *we* going? We're going to take a relatively short tour through the design of digital systems. When you are done with the course, you'll be able to design some such systems. Examples could be a washing-machine timer or a vending-machine controller.

Watch out! You'll know just enough to be dangerous! You are not going to get very deep into this design process—you can't in the time we have. But you will know the basics and you can build on those.

What are we going to do? This text is divided into nine chapters that pretty well cover the major aspects of digital systems and their design:

1. Introduction to designing digital systems with some examples, plus a description of some of the devices that we can use in our designs.

2. Numbers and arithmetic, where we will spend very little time. We do need to know a little about binary numbers and how they relate to decimal and perhaps hexadecimal,

as well as some simple binary arithmetic. This chapter, though, is more for reference later.

3.  No course on digital logic could be complete without some mention of Boolean algebra. This algebra is to digital systems what calculus is to mechanics. Yet we really don't need much formal Boolean algebra to design digital systems, just a few of the basic theorems and an understanding of how this all fits together.

4.  Combinational logic is the kind of logic where time is not a factor and there is no memory. Binary signals are combined to produce an immediate output. All digital systems will have combinational circuits within them.

5.  Building blocks combine into single chips many of the basic logic devices that we will use in designing combinational logic circuits. These blocks can be quite extensive, containing many circuit elements. These save us work when we are designing a system because they often will do on one chip what many basic chips could do.

6.  Sequential logic is where the fun is! These systems involve time and memory, remembering the past and acting upon both past events and current requests. These contain memory devices such as flip-flops. An example is the coin collector on a vending machine. You want it to record your 50% whether you put in two quarters or a quarter, two dimes, and a nickel. No combinational circuit can do that because it can't remember how much money has already been inserted.

7.  Registers and counters are particular instances of sequential circuits. They are so common in digital systems that there are standard building blocks that handle these functions. Counting the amount of money dropped into the coin slot of a vending machine is a good example of the use of a counter.

8.  Once we have all of this learning behind us, we will be ready to design finite state machines (FSM), which even sounds hairy! There are some pretty well-organized design techniques for doing this.

9.  Designers today don't design systems at the level we've been discussing. Instead, they work in a high-level hardware description language that takes much of the tedium out of detailed design. We'll finish up this text with an introduction to one of these languages, Verilog, and redo some of the earlier designs to demonstrate Verilog's use.

What could possibly be missing, you ask? Lots! We are not going to deal with hardware to any great extent. We aren't going to build physical circuits. We aren't going into any depth in any part of this because, after all, this is an introduction to the field, not a graduate design course. (LogicWorks simulations of all the examples are available on , and LogicWorks itself may be purchased from Amazon.com.)

So stick around for the ride—I think you'll enjoy it!

## 1.1    DESIGN—UP AND DOWN

When you start working a problem, one that doesn't have a single obvious solution, how do you start? I'm excluding problems such as a ball is dropped in a vacuum: how long does it take to reach a velocity of 2473 furlongs per fortnight? This problem has one answer and there are not very many ways of getting to it.

No, design problems generally don't have simple, fixed paths along which the design proceeds. They often involve going in the wrong direction, making the wrong decision, backing up and starting over, and so on. They generally have no single answer but rather can be solved lots of different ways.

Two approaches to design are *top–down* and *bottom–up*. The first involves looking at the big picture and breaking that picture up into smaller constituents. Each of these smaller parts is a complete entity, which will combine with all these other parts to make the finished whole.

But the smaller constituents may still be too large for one person or a small design team to handle. So we break them up again, getting even smaller constituents. Eventually, we'll have these constituents small enough that we can handle them, one by one, fairly easily.

More important, though, is that, if we are careful to break the problem at clear boundaries between subsystems, we will find that we have small constituents whose integrity we can be sure of. Then we can test each constituent completely, making certain that it does its job 100% correctly. When these are assembled into the larger system they were broken out from, the larger system should also work correctly.

How far must you go in breaking a problem down into these constituent parts? That depends (which probably isn't the answer you wanted to hear!). You break the system down until you have parts you can handle. You might perhaps choose to break the system into smaller parts than I would; maybe you would choose not to have so many smaller parts. It's really a judgment call based on how good you are at designing things, where the natural boundaries are, and what you and your team feel they can handle.

## 1.2    WHAT'S A DIGITAL SYSTEM?

Gee, I kind of answered that in the introduction to this chapter—a digital system is one that involves discrete quantities, most often binary digits. What we need to know now is what kinds of *things* we will find in digital systems.

The basic things are gates, which are devices that combine binary signals in fixed, simple ways. An example is a two-input AND gate. This gate accepts two binary signals on two separate inputs. It combines them in such a way that the binary output is 1 only if both inputs are also 1. We can combine gates into fancier combinational circuits such as the multiplexer. This device allows us to steer any one of a number of inputs to a single output.

The basic sequential device is the flip-flop, which is the simplest form of memory. The flip-flop *stores* a 1 or a 0. Applying the proper signals to its inputs causes the stored value to change—if it is a 1, it becomes a 0, for example. We can combine flip-flops into fancier sequential circuits such as the counter. This device counts the number of binary 1s presented to it over time on a particular input. Its "contents" are a binary number composed of several bits (binary digits).

As you probably now know, combinational circuits are made up of gates and fancier circuits that contain just gates. Sequential circuits are made up of both combinational circuits and devices that can "remember" such as flip-flops.

OK, that's enough words! Let's get into the design of digital systems. Well ... not so fast—we have lots of things to learn about. In the next couple of sections, I am going to design two simple systems, one combinational and one sequential. Why? Just to give you a feeling for what the process looks like.

## 1.3   COMBINATIONAL EXAMPLE

Suppose that an automobile has three switches that can operate the turn signals. Two of these are on the stalk behind the steering wheel, one for the left turns and one for the right. The third is the emergency-flasher button that activates both the left and right signals.

Design the digital logic needed to activate the left- and right-turn/emergency signals.

Oh, oh! I don't have complete information. For example, what kind of switches are these? How do they provide the information, in binary, presumably, that they have been actuated? But when I am developing something, I often find that the statement of the problem is far from complete.

What I need are the *specifications* that tie down the details so completely that I'm not going to make wrong assumptions. These specifications also will serve as a guide and a contract, because once I have everything tied down, I will get my customer to agree to them. Then both the customer and I will know when I have finished the design. The specifications form a major part of the contract between us. (Don't forget to include some money considerations too!)

### 1.3.1   Specifications

OK, here are what I am proposing as the specifications:

- The three switches are arranged with appropriate logic so that they produce a binary 1 when actuated and a binary 0 otherwise.

- The left- and right-turn switches cannot be actuated at the same time because they are mechanically separate on the stalk.

- Operating the turn signal is to override the emergency flasher.

- The design is only of the activation of the lights; flashing them is not a part of the design.

There! Does that tie things down? Look back through the information and see if you can find anything missing. If you were my customer, you'd have to do this, even if you weren't an expert on digital logic. Once you agree to these specs, you are bound to pay me for my work when I show that I have done exactly what they say. Mistakes can be expensive!

The next step is to say this all again in a logic statement.

## 1.3.2  Truth Table

The truth table is a complete tabulation of all the possible outcomes for every possible arrangement of inputs. In effect, it "tells the truth, the whole truth, and nothing but the truth."

To develop the truth table, I'll need names or labels on the input signals from the switches and the output signals to the lamps. I'll call the left-turn switch L; when $L = 0$ the switch is not actuated, and when $L = 1$ the switch is actuated. The right-turn switch will be R and the emergency-flasher switch will be F.

Output signals will go from my logic to the lamps at the front and rear of the car. I'll call the signal to the left bulbs LT; when $LT = 0$ the bulbs are not on, and when $LT = 1$ the bulbs are to be on. The right bulbs will be RT.

Now for the truth table, shown in Fig. 1.1-1. Notice the headings are the input and output signals. On the left you'll see all eight possible combinations of three binary digits. I

| L | R | F | LT | RT |
|---|---|---|----|----|
| 0 | 0 | 0 | 0  | 0  |
| 0 | 0 | 1 | 1  | 1  |
| 0 | 1 | 0 | 0  | 1  |
| 0 | 1 | 1 | 0  | 1  |
| 1 | 0 | 0 | 1  | 0  |
| 1 | 0 | 1 | 1  | 0  |
| 1 | 1 | 0 | x  | x  |
| 1 | 1 | 1 | x  | x  |

FIGURE 1.1-1: Truth table for turn signals.

have them in ascending numerical sequence. On the right are the possible outcomes of each of the eight input combinations.

Let's look specifically at each line so that you can see how I derived the outputs from the specifications:

000   Both outputs are 0 because no switch is actuated.

001   The emergency flashers are being requested, so both bulbs are lit.

010   This is a request for a right turn.

011   Both a right turn and the emergency flashers are being requested, but the specifications say that the turn signal is to dominate.

100   This is a request for a left turn.

101   A left turn dominates when both a left turn and the flasher are requested.

110   Here the output is shown as *x*, which means that I don't care what the output actually is. Why? The specs say that both L and R cannot be actuated at the same time, so the output, whatever it might be, will never happen. (The *x* is called a *don't care.*)

111   Here's another impossible situation.

All eight possible switch combinations have been listed and the outputs specified. I check these against the specifications to make sure I'm still on the right track.

Now I need to reduce this logic representation to a logic statement that can lead to logic gates.

### 1.3.3   Reduction and Simplification

The technique I will use to get the logic statements is called the *Karnaugh Map*, which is a graphical way of doing the job. This is just one way that I could do the job.

Figure 1.1-2 shows the Karnaugh maps. I will do this without explanation, but I'll bet that if you look at the maps closely you'll see how I got them. The less obvious part is how I

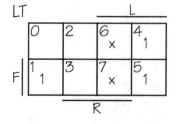 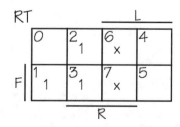

FIGURE 1.1-2: Karnaugh map for turn signals.

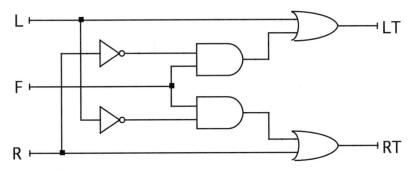

**FIGURE 1.1-3:** Combinational logic for turn signals.

get the following logic equations from them:

$$LT = L + F \bullet \overline{R}$$
$$RT = R + F \bullet \overline{L}$$

Actually, the equations are quite readable if you know that the + means *or*, the • means *and*, and the bar over a symbol means to take the logical complement (i.e., if $L = 1$, L with a bar over it is 0, and vice versa).

Read the equation for LT: the left-turn bulb is to light if the left-turn switch (L) is actuated, or if the emergency-flasher switch (F) is actuated and the right-turn switch (R) is not actuated. Try this for the other equation.

But the real test of all this is to implement it.

### 1.3.4 Implementation

Figure 1.1-3 is a drawing of the entire circuit. The circuit is organized in exactly the same manner as the equations. There are only six parts and just three different kinds:

- Inverter, the triangle with the open dot (the "donut") at its point. The output of the inverter is the logical complement of the input—if a 1 comes in, 0 comes out, and vice versa.

- *And* gate, a straight back and a round tummy. The output will be a 1 only if both inputs are 1.

- *Or* gate, a curved back and a pointy tummy. The output will be a 1 if any input is a 1 (including both inputs).

I use LogicWorks to simulate this circuit's operation and show that it does indeed perform according to the specification.

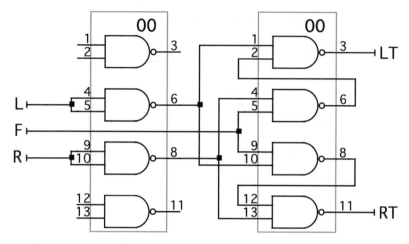

**FIGURE 1.1-4:** Circuit using standard parts.

### 1.3.5   Payment!

I will demonstrate my design to my customer, show that it meets the specifications, and then turn over to the customer my design, probably my simulation file, and the documentation on the circuit, which will show not only the results of my design but my choice of standard parts to use in building the circuit. Figure 1.1-4 shows the circuit made of commercially-available 7400-series parts.

Payment time, please!

## 1.4   SEQUENTIAL EXAMPLE

Now let's make the lamps flash. We want them to flash at a reasonable rate when activated. Again I'll start with the specifications.

### 1.4.1   Specifications

Either lamp is to flash at an appropriate rate when the signal to activate the lamp is received. The specifications are rather simple:

- The LT and RT signals from the previous design are to be the inputs to the new circuit.
- The lamps are to flash at the rate of one flash every two seconds. A clock signal (Clk) will be available that "ticks" every second.
- The outputs to the light bulbs are to be LB and RB; bulbs light when the signal is 1.

Notice that the clock is an external signal that provides the timing. There are circuits that generate clock ticks but I'm not going to worry about them now.

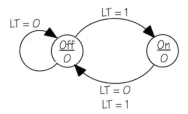

FIGURE 1.2-1: State diagram: LT flasher.

I will develop the design for the left-turn bulb; the design for the right-turn bulb will be identical except for signal names.

### 1.4.2   State Diagram

The truth table was one way to describe the combinational circuit in complete detail. The *state diagram* is a way to do this for a sequential circuit. This diagram shows the transitions from one state to another. In my design, one state is with the bulb lit and the other state is with the bulb dark. These two states will appear in my state diagram.

The state diagram for the left-turn lamp is shown in Fig. 1.2-1. There are three things to notice:

- The input to this diagram is the left-turn signal from the combinational circuit I just designed. It always appears as LT written along the lines that show the transitions between states.
- The circles are the states: the name of the state is above the line and the binary output is below the line. The "Off" state has an output of $LB = 0$, the "On" state has an output of $LB = 1$.
- A transition along one of the paths is made *only* when the clock tick comes along. This is not shown specifically in the diagram but is assumed, since this is a clocked circuit.

This diagram is not hard to read. Suppose, for example, that the circuit is in the "Off" state, which means that the most recent tick of the clock left it there. Suppose that the input $LT = 0$. At the next tick of the clock, the state will not change because the transition arrow for $LT = 0$ ends on the "Off" state.

Suppose that the circuit is still in the "Off" state and the input is now $LT = 1$. At the next clock tick, the state will change to the "On" state.

Follow the rest of the transitions to see if I've done it right!

| Current state | Input LT | Next state |
|---------------|----------|------------|
| Off | 0 | Off |
| Off | 1 | On |
| On | 0 | Off |
| On | 1 | Off |

FIGURE 1.2-2: State table.

This information can also be expressed via a State Table, as shown in Fig. 1.2-2. Some designers use diagrams, some use tables, but most use a combination of these. Fig01-04-01.eps

Now I need to reduce this to a more "binary" design.

### 1.4.3   Transition/Excitation Table

The State Diagram or the State Table can be translated rather easily into the Transition Table, as shown in Fig. 1.2-3. The first three columns show the same transitions that I showed in the previous table. For example, the first line says that if the current state is 0 (i.e., the left-hand circle) and the input is 0 (i.e., $LT = 0$), then the next state is 0.

From this table I will derive the excitations for the flip-flop that I choose. Right now this is beyond us; we'll do it in detail in Chapter 6. The Next State column is also the excitation (input) for a D-type flip-flop. The last column is the excitations for a JK-type flip-flop.

From these, the derivation of the logic functions for the inputs to the appropriate flip-flop is easy to do (once we've gotten that far in the course!). For the D flip-flop, the excitation is

$$D = \overline{Q} \bullet LT$$

and for the JK flip-flop, the excitations are

$$J = LT, K = 1.$$

| Current state Q | Input LT | Next state Q | Excitation J | K |
|-----------------|----------|--------------|--------------|---|
| 0 | 0 | 0 | 0 | x |
| 0 | 1 | 1 | 1 | x |
| 1 | 0 | 0 | x | 1 |
| 1 | 1 | 0 | x | 1 |
| | | D | | |

FIGURE 1.2-3: Transition/excitation table.

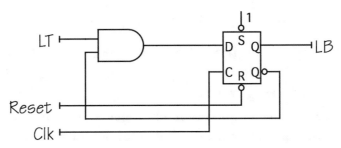

FIGURE 1.2-4:  D flip-flop implementation.

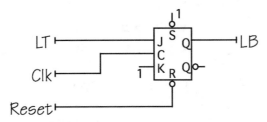

FIGURE 1.2-5:  JK implementation.

(These equations are for the left-turn signals; the equations for the right turn are the same.)

### 1.4.4   Implementation

Implementàtion using a D flip-flop is shown in Fig. 1.2-4; the JK form is in Fig. 1.2-5. We'll look at a complete simulation, including the combinational logic, later.

### 1.4.5   Payment!

I'm done, and I say that it meets specs, so I am due the money for my work. Again, I'll turn over to the customer my detailed design, my reduction of this circuit to commercial parts, and my documentation.

## 1.5    DOCUMENTATION

Something most engineers detest doing is documenting their work. But without the documen-tation their designs are almost useless. There is simply nothing to communicate the functioning of the design to others. That absence makes it hard to sell a system, repair it, modify it later, interface it with other systems, and so on.

Documentation has to accomplish at least three things:

1. Present the design itself on paper so that the user can know what has been done, consider how to repair or modify it, and develop additional uses. It is generally very difficult to determine a circuit configuration by "reverse engineering" the physical circuit.

2. Explain the interfacing of the system. Just about every design must work in concert with other parts of a larger system. A detailed description of the interfaces for the design is required for the successful interconnection of the design with its neighbors. The interface description also gives a way of testing the design itself to see that it meets the specifications before embedding it into a larger system.

3. Record for posterity the design information to support such actions as patent applications, defense against patent litigation, and so on.

In other words, ya gotta write the thing up! On top of the technical needs for such write-ups, the quality of your work may be judged by the quality of your documentation, especially by managers a level or two above you. So there's value to you in this process.

Much of your documentation will depend upon symbolic drawings and descriptions of signals. Let's take a look at these in a little more detail. But even though I am writing a fair number of words about these, I think you'll learn at least some of this by example and by doing.

### 1.5.1   Logic Symbols

You've already seen some logic symbols. These can be classed as the "traditional" symbols, mainly because they have been used by several generations of engineers. (I first encountered them as an undergraduate in the 1950s.) They make use of their shape as well as labels to indicate the functions they perform.

Figure 1.3-1 shows a number of these traditional symbols. You'll understand more about their functions as we go further into the design process, but it's interesting to see some of the things that we can use.

*And* gates produce an output of 1 only when all of the inputs are 1s.

*Or* gates produce an output of 1 when any combination of inputs is 1s.

*Exclusive or* gates are the same as *or* gates except that the "both" case is not included. That is, the output is 1 if either input is a 1 but not both. The plain *or* gate is sometimes called the *inclusive or* to help distinguish the two.

*Buffers* are like *not* gates without the inversion. Their purpose is amplification.

*Nand* gates are the same as *and* gates but the output is inverted. Hence the output is 0 only when all of the inputs are 1s.

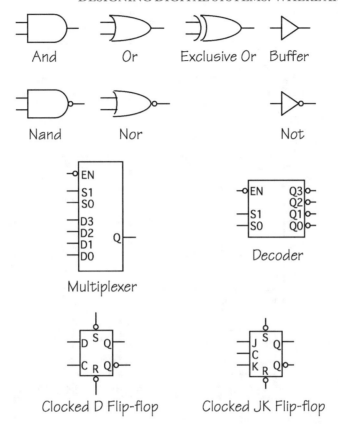

**FIGURE 1.3-1:** Traditional symbols.

*Nor* gates are the same as *or* gates but the output is inverted. Hence, the output is 0 when any combination of inputs is 1s.

*Not* gates are just inverters, changing an incoming 1 to a 0 and vice versa. The *not* gate often also serves as a signal amplifier, strengthening the signal so that it can provide inputs to a large number of other gates or drive some devices such as LEDs and small relay switches.

*Multiplexers* select one input from many (here from one of the inputs D0-D3) and present it on the output Y.

*Decoders* assert only one output (here Y0–Y3) depending on the selection inputs (here S0–S1).

*Flip-flops* come in many forms, two of which you've seen in my example earlier. The clocked D flip-flop changes its state (either 0 or 1) only when the clock tick comes along. The new state will be the value on the D input. The JK flip-flop is more complicated and we'll see how it works later in Chapter 6.

*Dots* or *bubbles* appear with a number of these symbols. In every case, they mean that the signal is inverted as it passes through the bubble. This means that a 1 would become a 0 and

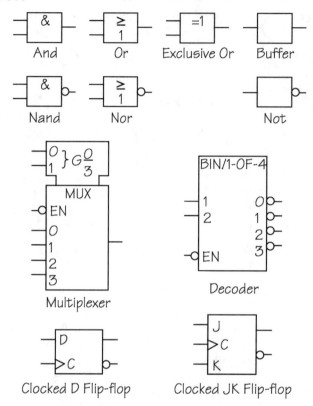

**FIGURE 1.3-2:** IEEE standard symbols.

vice versa. It's just like having a *not* gate there. Note the bubbles on the decoder outputs, meaning that the asserted value of the outputs are 0, not 1.

These traditional symbols are standard, and there is an IEEE published standard (IEEE Std 91-1984) that tells how the shapes are to be drawn. You should try to draw your symbols in reasonable conformity to these.

A note of caution: these symbols are fairly uniformly recognized, so it would be poor practice to use one of the shapes for something non-standard.

### 1.5.2   IEEE Standard Symbols

In 1984, the IEEE developed a set of standard symbols that provides a richer collection of different logic functions (IEEE Std 91-1984). They are mostly rectangles rather than traditional shapes. Different kinds of blocks and the symbols inside them describe the logic functions of the symbols.

*And* is indicated by the ampersand &.

*Or* produces an output of 1 if at least ($\geq 1$) one input is 1.

*Exclusive or*, on the other hand, required exactly one ($=1$) input to be 1 for an output of 1.

*Buffer* strengthens the signal.

*Nand* is an *and* gate with a complemented output.

*Nor* is an *or* gate with a complemented output.

*Not* complements the signal and strengthens it.

*Multiplexer* has two parts, the selection logic, shown at the top in the T-shaped box, and the function itself. If a 0 is placed on the enable (EN) line, the output on the right is connected to the input that has been selected by the two inputs at the top.

*Decoder* asserts the output selected by the two input lines if the enable (EN) line is 0. An asserted output carries a 0 because of the bubble.

*Flip-flop* diagrams are about the same in both symbol standards.

*Dots* or *bubbles* work the same way in both standards.

Which standard should you use? One answer should be obvious in this text—I prefer the traditional symbols because their shapes help when I'm reading diagrams. But some employers require use of the IEEE symbols by their employees and contractors. Right now, though, the traditional symbols seem to be much more popular.

## 1.5.3   Wires and Buses

Wires are pretty simple to draw, just lines from outputs to inputs. When we are drawing logic devices, we generally omit the power wiring, which are usually two wires to specific pins on the logic chips. Including those clutters up the drawing and makes it hard to follow the logic itself. We also omit ground wires, although logic signals are generally referenced to ground.

Sometimes we have many wires that group in some logical way. An example is the eight-bit data bus of a microprocessor. Let's consider an example of a bus. Suppose we have an address bus of eight bits with wires named A0, A1 ..., A7. We want two lines of this bus to feed two inputs of an 8-to-1 multiplexer.

Figure 1.3-3 shows how this is done without drawing all the lines. The eight signals have been created by logic that is off to the left in the diagram. Individual signals are broken out of the bus as needed.

## 1.5.4   Signal Names and Levels

Name signals in a meaningful way! The reader is helped a lot when you name a select signal as "SELECT1" rather than "B" or "TAHITI." Meaningful names also help when it comes to documenting how your logic system works.

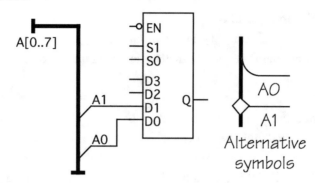

FIGURE 1.3-3:  Bus symbols.

Pump_on ▷∘ /Pump_on

FIGURE 1.3-4:  Negation

Signals also have levels associated to them. For example, the signal SELECT1 may be active when it is 1. (We say that the signal is *asserted* when it is a 1 and that it is *active-high*.) But our signals don't have to be 1 to be active. We can perfectly well choose, for a particular signal, 0 as the active level.

A signal is assumed to be active when it is a 1 unless it is specially marked. So, a signal such as SELECT1 is assumed to be active-high.

If we want a signal to be active-low, we generally must mark it in some manner to show the reader our choice. There are several ways of doing this. Suppose the signal SELECT2 is to be active-low, which means that we wish to "select" something when this signal is 0. We indicate that by writing SELECT2' or SELECT2* or /SELECT2 or by placing a bar over the signal name.

Figure 1.3-4 shows a signal that is complemented. Note that its name has been changed by the slash symbol. (When I write Boolean logic, I'll generally use the prime mark as in SELECT2'; for diagrams I'll use the slash as in /SELECT2.)

## 1.6    LOGIC CIRCUIT FAMILIES

Logic circuits come in families. Well, no, not the usual collection of grandparents and moms and dads and kiddies, all of whom may or not be compatible. The members of a logic family are compatible to the $n$th degree, which is what the family is all about.

All the members of a logic family have very similar characteristics. All require the same power supply voltages. All respond to the same logic signal levels. All produce the same signal levels. All of them operate at similar speeds.

The members of a logic family are designed to work together simply. In most cases, a signal is moved from one member to another "just by running wires." You don't have to pay attention to signal levels, voltages, or drive currents. You do, however, have to consider timing, because a signal that arrives with too much delay may not cause the desired operation.

There are several popular families, more than I want to even mention here. Two, however, are quite common because they are simple, cheap, and versatile. (They are also fairly forgiving of wiring errors—it takes some effort to accidentally let the magic smoke out of a chip!) Let's look at two of these families.

### 1.6.1   TTL Family

The transistor–transistor logic family (TTL) has been around for over 40 years. The two Ts in the name tell a little about the physical arrangement of the internal components. Part labels generally begin with a number, either 74 (commercial grade parts) or 54 (military grade).

Within the TTL family are several subfamilies, separated by such characteristics as power consumption and speed. The particular subfamily is designated by letters after the initial number. Two examples are

74LS_ series (low-power Schottky), which are simple, require relatively small amounts of power, and operate at moderate speed, fast enough for many applications (such as vending-machine logic).

74F_ series (fast), which are faster than the 74_LS series. These, and others that are even faster, may found in such systems as printers.

The blank after the letters is filled in with two to four digits. But the digits are the same for standard circuits. For example, 74LS_00 and 74F_00 are both chips containing four two-input Nand gates.

The TTL family typically requires a 5-volt power supply.

### 1.6.2   CMOS Family

The CMOS logic family (complementary metal-oxide semiconductor) is a very low-power family of chips. Typical power consumption when the chip isn't doing anything is in the microwatt range. The members of this family require essentially zero power except during switching, which means that their average power consumption depends on how many times the output switches per second. In other words, power consumption is proportional to frequency.

I'll mention just three subfamilies here:

40_ series, which was the original family. Its devices are moderately fast.

74HC_ series (high-speed CMOS), which is popular and has many different types of devices available. But it, like the 40_ series, will not interface easily with the members of the TTL family. Yet we sometimes need to do this because the CMOS families do not have as many varieties of logic devices, something I am not going to cover in this text.

74HCT_ (high-speed CMOS, TTL compatible), which resolves that problem. But here too, the variety is not very extensive.

As with the TTL families, standard numbering is used. So a 4000, a 74HC_00, and a 74HCT_00 all have four Nand gates on one chip.

CMOS has another advantage—lower power supply voltages. While 5 volts is still common, 3.3 volts, 3 volts, and even less than 3 volts are available.

So where does this get us? Just the facts, just the facts. What I've just covered is mostly background information so that, when you encounter the chips of the real world, you won't be totally illiterate!

## 1.7   SUMMARY

Well, we've gotten this all started and you've seen three logic circuits. After looking into what digital systems are, we designed a combinational circuit for the turn signals. We started with an incompletely-specified word problem, produced specifications, reduced these to a truth table, and created a logic circuit to do the job.

Combinational circuits have no memory—sequential circuits do. We added a flasher to the turn signals, proceeding from the specifications to the State Diagram/Table to the Transition/Excitation Table and finally to the logic circuit, which contained flip-flops as the memory elements.

And what's next? A brief look at numbers and arithmetic, mostly in binary, and codes. But only briefly.

CHAPTER 2

# Numbers and Arithmetic: Counting 101

Numbers are everywhere. We count money, we calculate, we inventory, we do arithmetic. In fact, we take numbers for granted, something that has existed since the dawn of time.

But numbers had to be invented just like lots of other things. Fortunately, no one back then took out a patent or we'd be paying royalties every time we go to the grocery store.

Of course, the invention that we are most familiar with is decimal numbers, base 10, all the fingers and thumbs. Computers don't do well with decimal, at least not directly. So when we deal with numbers in digital logic, it's much more likely that we'll use binary, base 2.

That's mostly what this chapter will be about—binary. And because hexadecimal is useful and common, we'll throw in a little of that too. But let's start by looking at decimal—an unpatented number system.

## 2.1    DECIMAL

Decimal numbers are based on tens. But these numbers didn't just happen at the beginning of time. Somebody or some group developed them over a long period of time as there was more need for numbers and counting, and finally arithmetic.

Consider first a simple prehistoric tribe living somewhere. Each family had a cow, so the concept of "one" was well known. Some larger families needed two cows to sustain them, so the concept of "two" was probably also well known. But as you might expect, certain well-to-do families had many cows. Hence the concept of "many" was most likely also known. So this culture had a very simple counting system: one–two–many.

The Romans had an interesting number system to keep track of larger herds and such, the so-called Roman numerals. Roman numerals were fashioned after simple tallies (Fig. 2.1-1). "V" was a handful, "X" was two hands-full, and so on.

The Romans didn't really have the concept of zero, though. Oh, sure, they knew about "nothing," because they could certainly sit around the baths and lament that the city had no good gladiators any more. But the number system did not have a symbol for zero.

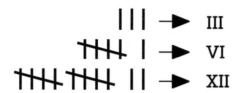

FIGURE 2.1-1: Roman-numeral tallies.

The Romans also didn't know about position within a number system. For example, consider the number

$$LXXVIII$$

You can readily translate this into "78" in our common number system. But let's look more closely. There are two "X" symbols in the number. They are in the 6th and 5th positions from the right. But they both mean "ten" regardless of position. The "I" symbol means "one." Position doesn't matter.

Our number system is decimal, and position does count. Somebody, long, long ago, came up with the concept of "zero." This gave the system a placeholder so that a numeral in one position has a different value from a numeral in another position.

An example of this is the number

$$20629$$

Here, the numeral symbol "2" appears twice. But each carries a different value. In the position second from the right, it has the value "twenty"; in the position fifth from the right, it has the value "twenty thousand."

The concept of the point is also important. The point tells us where to begin counting positions. In the previous example, the point was implied: it's on the right after the 9. Consider the number

$$32.063$$

The symbol "3" appears twice. In the position second to the left of the point, it has the value "thirty"; in the position third to the right of the point, it has the value "three thousandths."

Each position has a unique multiplier value based on its position relative to the point. The position just to the left of the point has a multiplier value "1" or $10^0$. The next position to the left has a multiplier value "10" or $10^1$. Working to the right, the first position to the right of the point has a multiplier value "0.1" or $10^{-1}$. And so on. We don't usually think of this, at least in this way, but it will be important when we work with binary numbers.

So symbol and position both count. Since the base of the decimal number system is 10, the positions are powers of 10. Symbols take on numerical values that are computed by multiplying the symbol value by the positional value of the power of ten. That's important, because it applies to numbers written in any base, which leads us to binary.

## 2.2    BINARY

Binary numbers use only two symbols, the numerals 0 and 1. The 0 is still the position holder; the only "value" symbol is therefore 1. The base of this number system is 2, so all of the positions in the system are powers of 2.

Consider the binary number

$$110100.1011$$

I can "decode" this number into decimal by writing it with powers of 2 in the right positions:

$$\begin{aligned}110100.1011 &= 1 \times 2^5 + 1 \times 2^4 + 0 \times 2^3 + 1 \times 2^2 + 0 \times 2^1 + 0 \times 2^0 \\ &\quad + 1 \times 2^{-1} + 0 \times 2^{-2} + 1 \times 2^{-3} + 1 \times 2^{-4} \\ &= 32 + 16 + 4 + 1/2 + 1/8 + 1/16 \\ &= 52^{11}/_{16} = 52.6875 \end{aligned}$$

So a simple way to get from binary to decimal is to write the powers of 2 and then finish the computation.

But notice that I did all of the arithmetic in *decimal* because that's the number system I know best. If I were a computer, though, working with binary numbers and using binary arithmetic, I'd have to do all the computations in base 2.

Now we know a way of getting from binary back to decimal. How about going from decimal to binary? There are several ways of doing this, but I'll show only one, successive divisions or multiplications by 2.

This algorithm for conversion from decimal to binary requires that I separate the number into its integer and fraction components. Each is treated differently.

Let's start with the integer part of the last example, the 52. Figure 2.2-1 shows the algorithm, successive divisions by 2. The remainders become the binary digits, but be careful to read them from *bottom to top!*

The fraction part requires successive multiplications by 2 (Fig. 2.2-2), chopping off the digit that ends up to the left of the point and using that as the binary digit. Here we read from the *top down.*

$$
\begin{array}{ll}
2\,\underline{|\,52} & \bullet \\
2\,\underline{|\,26} & 0 \\
2\,\underline{|\,13} & 0 \\
2\,\underline{|\ \ 6} & 1 \\
2\,\underline{|\ \ 3} & 0 \\
2\,\underline{|\ \ 1} & 1 \\
2\,\underline{|\ \ 0} & 1 \\
\end{array}
\quad = 110100.
$$

FIGURE 2.2-1:  Integer conversion.

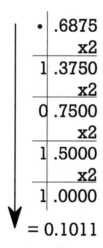

$$
\begin{array}{rl}
\bullet & .6875 \\
 & \text{x2} \\
\hline
1 & .3750 \\
 & \text{x2} \\
\hline
0 & .7500 \\
 & \text{x2} \\
\hline
1 & .5000 \\
 & \text{x2} \\
\hline
1 & .0000 \\
\end{array}
$$

$= 0.1011$

FIGURE 2.2-2:  Fraction conversion.

The final result is the two parts combined. So the conversion yields

$$52.6875_{10} = 110100.1011_2$$

I'm hiding one important fact from you by my choice of example, though. While all decimal *integers* have exact binary representations, almost no decimal *fraction* comes out exactly—unless you include an infinite number of binary digits.

In other words, decimal fractions don't convert exactly to binary; this is a source of error when using a computer to handle decimal data.

How do you know how far to go when the multiply-by-2 scheme doesn't terminate? Recall, perhaps from physics, that a decimal fraction such as 0.6875 implies a certain precision, here a precision of one part in 10,000. Conversion to binary should have about the same precision. This means that there should be about 3.3 times as many binary digits in a fraction as there are decimal digits in the equivalent fraction. I'll leave the why of this to you. (Yes,

I know, I remember that as a student I hated statements like that! It usually means that the professor hasn't figured it out yet.)

## 2.3    HEXADECIMAL

Hexadecimal numbers use a base of 16, which perhaps seems like a strange choice. But they are convenient for human use when we want to represent binary numbers but don't like to write all those zeros and ones. You'll see why at the end of this section.

Hexadecimal numbers need 16 symbols. Convention says we are to use the first six letters of the alphabet as the digits beyond the first ten. So the symbols for hexadecimal are

$$0\ 1\ 2\ 3\ 4\ 5\ 6\ 7\ 8\ 9\ A\ B\ C\ D\ E\ F$$

where A has the value 10 in decimal, B is 11, and so on.

The conversion algorithms are the same as in binary, but instead of using 2, use 16.

Consider the hexadecimal number 2A3.1D The conversion uses position and powers of 16:

$$\begin{aligned}
2A3.1D &= 2 \times 16^2 + A \times 16^1 + 3 \times 16^0 + 1 \times 16^{-1} + D \times 16^{-2} \\
&= 2 \times 256 + 10 \times 16 + 3 \times 1 + 1 \times 1/16 + 13 \times 1/256 \\
&= 675^{29}/_{256} = 675.11328125_{10}
\end{aligned}$$

As with binary, the number of digits to keep for equivalent precision is a question. In the fraction, keep about 1.2 times as many decimal digits as there were in the hexadecimal fraction. (Aha, another problem assignment!)

Going from decimal to hexadecimal uses the same algorithms as binary, except that division and multiplication are by 16. As before, we have to separate the integer and the fractional parts first.

Then successive divisions of the integer by 16 (Fig. 2.3-1) and successive multiplications of the fraction by 16 (Fig. 2.3-2) yield

$$675.11328_{10} = 2A3.1CFF\ldots_{16}$$

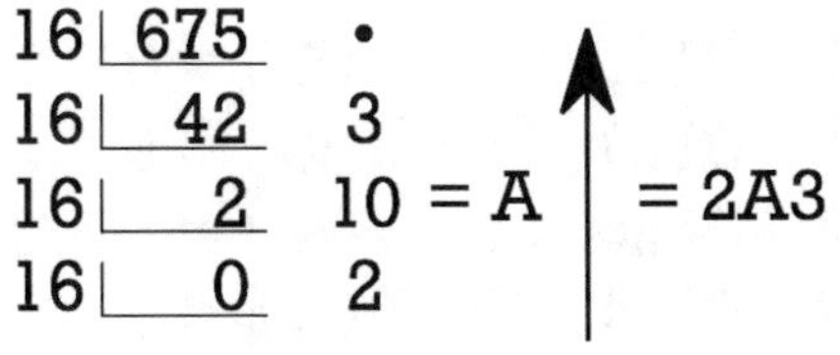

$$
\begin{array}{lll}
16\underline{|\ 675} & \quad\cdot \\
16\underline{|\ \ \ 42} & \quad 3 \\
16\underline{|\ \ \ \ \ 2} & \quad 10 = A & = 2A3 \\
16\underline{|\ \ \ \ \ 0} & \quad 2
\end{array}
$$

FIGURE 2.3-1: Hex integer conversion.

$$
\begin{array}{r|l}
\bullet & .11328 \\
\hline
 & \text{x16} \\
\hline
1 & .81248 \\
\hline
 & \text{x16} \\
\hline
12 & .99968 \\
\hline
 & \text{x16} \\
\hline
15 & .99488 \\
\hline
 & \text{x16} \\
\hline
15 & .91808 \\
\end{array}
$$

etc.

$$= 0.1(12)(15)(15)...$$
$$= 0.1CFF...$$

FIGURE 2.3-2: Hex fraction conversion.

There's another property of hexadecimal that makes it useful for humans when they are working with computers and binary numbers. If the base of one number system is an integer power of the base of another number system, conversion can be done by inspection.

In our case, binary, which is base 2, and hexadecimal, which is base 16, are related by $2^4 = 16$. This says that groups of four binary digits translate directly into a single hexadecimal digit. Figure 2.3-3 is the table of these translations.

Let's translate to hexadecimal the binary number we used in Section 2.2. Group the binary digits by fours starting at the point and working outward. Fill in leading or trailing zeros as needed. Then translate.

$$110100.1011_2 = 0011|0100| \, . \; 1011_2$$
$$3 \qquad 4 \quad . \quad B = 34.B_{16}$$

| | | | |
|---|---|---|---|
| 0000 = 0 | 0100 = 4 | 1000 = 8 | 1100 = C |
| 0001 = 1 | 0101 = 5 | 1001 = 9 | 1101 = D |
| 0010 = 2 | 0110 = 6 | 1010 = A | 1110 = E |
| 0011 = 3 | 0111 = 7 | 1011 = B | 1111 = F |

FIGURE 2.3-3: Binary to hex.

Going the other way requires that we write four binary digits for each hexadecimal digit:

$$2A3.1D_{16} = 0010|1010|0011|.0001|1101$$
$$= 1010100011.00011101_2$$

Keep in mind that computers deal just with binary internally—this hex thing is to make the binary numbers easier for us to write.

## 2.4    BINARY ARITHMETIC

Binary arithmetic is pretty simple, which is one of the reasons why computers do it (i.e., computers are pretty simple-minded, or are they symbol-minded?). Memorizing the addition and multiplication tables isn't too difficult (Fig. 2.4-1).

Doing arithmetic in binary isn't all that hard, either, although it takes lots of binary digits to get much of a number. (Remember the 3.3 factor—it takes about 3.3 binary digits for every decimal digit in a value.)

Let's add two binary numbers, here 53 + 422 (Fig. 2.4-2). Notice that the carries work just as they do in decimal.

Multiplication isn't hard, either (Fig. 2.4-3). If the multiplier is a 1, copy down the multiplicand (the top number). If the multiplier is a 0, copy down zeros and then add.

But . . . as usual things aren't going to stay that simple. We want to do arithmetic using hardware. In other words, we want to design logic circuits that will do arithmetic. But there has to be some place to hold the numbers we are working with. This implies registers, which are hardware. Hardware isn't infinite! So the registers have limited length. That's where binary arithmetic gets interesting.

| + | 0 | 1 | | x | 0 | 1 |
|---|---|---|---|---|---|---|
| 0 | 0 | 1 | | 0 | 0 | 0 |
| 1 | 1 | $^1$0 | | 1 | 0 | 1 |

FIGURE 2.4-1: Binary + and x.

```
   110101
+   10110
 1001011
```

FIGURE 2.4-2: Binary addtion.

$$
\begin{array}{r}
110101 \\
\text{x} \quad 110 \\
\hline
000000 \\
110101 \\
110101 \\
\hline
100111110
\end{array}
$$

FIGURE 2.4-3: Binary multiplication.

We are going to take a look at binary arithmetic, both addition and subtraction in a restricted-length environment. Subtraction will bring up complement arithmetic.

### 2.4.1    Binary Addition

Addition when the number of bits is restricted means that some results won't fit. Let's keep this simple by restricting our numbers to four binary digits. In other words, let's suppose that our hardware registers and all our logic circuits can handle only four such digits. (Modern microprocessors have register lengths of 8, 16, 32, and 64 bits.)

Figure 2.4-4 shows two addition examples using exactly four bits. In the example on the left, $9 + 5 = 14$, which fits into four bit positions. This addition worked.

In the example on the right, $11 + 6 = 17$. But 17 won't fit into four bits, so the result is an *overflow* off the left end of the computation. The result appears to be 1 with an overflow, which is just what it is. How the overflow is processed depends on what you are trying to do.

### 2.4.2    Subtraction and 2's-Complement

Subtraction brings up an interesting new problem (Fig. 2.4-5). The subtraction on the left, $9 - 5 = 4$, works fine when I borrow properly (remember "borrowing" in about the third grade?).

The subtraction on the right, $5 - 9 = ?$, runs into a problem. We know that the result should be $-4$, but how is a negative number represented?

$$
\begin{array}{r}
1001 \\
+\ 0101 \\
\hline
1110
\end{array}
\qquad\qquad
\begin{array}{r}
1011 \\
+\ 0110 \\
\hline
1\,|\,0001
\end{array}
$$

FIGURE 2.4-4: Restricted addition.

$$\begin{array}{r} 1\!\!\!\diagdown\!\!1001 \\ -\ 0101 \\ \hline 0100 \end{array} \qquad \begin{array}{r} 0101 \\ -\ 1001 \\ \hline ?100 \end{array}$$

FIGURE 2.4-5: Restricted substraction.

Negative numbers can be represented in several ways, but one thing is obvious: we need a sign to indicate which numbers are positive and which are negative. So I will make a simple change to my four-bit number system—I will make the leftmost bit the *sign bit*. If this bit is a 0, the number is +, and if it is a 1, the number is −.

This signing scheme uses up one of the four bit positions, so my numbers now have only three magnitude bits. Oh, you say, why not extend the number to five bits? Well, recall that I said my hardware restricts numbers to four bits, so I'm stuck at four.

This is called *signed magnitude*, which means that the number has a sign and then the rest of the bits represent its magnitude. Here are two examples:

$$0101 = +5 \quad 1101 = -5$$

This is the same scheme that you and I use all the time in decimal.

But computing with signed-magnitude numbers isn't the simplest way to do things. A very common scheme is to use *twos-complement* representation for negative numbers:

$$0101 = +5 \quad 1011 = -5$$

The number on the left is +5, with the sign bit (leftmost bit) 0 and the rest of the bits (101) giving the magnitude.

The number on the right is a twos-complement number. I know this because I have said I'm using that scheme and because the sign bit on the left is 1, indicating a negative number. But the magnitude bits are the *complement* of the magnitude of 5. This is found by subtracting the original magnitude from the next higher power of 2 (Fig. 2.4-6).

$$\begin{array}{r} 1000 \\ -\ 101 \\ \hline 011 \end{array}$$

FIGURE 2.4-6: Complementing.

Note that this scheme also changes the available range of values that can be covered by four bits. If all four bits are magnitude bits, then numbers can range from $0000 = 0$ to $1111 = 15$. If instead we use 2's-complement, the range is from $1000 = -8$ to $0111 = +7$. There are still 16 possible values, but they cover both negative and positive numbers.

Overflow gets interesting in this twos-complement scheme! Overflow requires a look at the sign of the result. For example, if adding two positive numbers creates a negative number, there's clearly a problem! I won't solve that here.

## 2.5   CODES

Binary numbers aren't always meant to be numerical values. Instead, they might be numerical codes for various things. An example of such a code is ASCII, a code for representing in binary the various characters that we use for such things as writing books.

Two codes of interest are Binary-Coded Decimal (BCD) and Gray codes.

Sometimes we want to represent decimal numbers without converting them to binary values using the algorithms we've just seen. In other words, we want to work with the actual decimal values, even though the representation must be in the form of bits.

The most common way of encoding decimal digits in binary is the BCD code (Fig. 2.5-1). Notice that this is just the first ten values that I listed in the hexadecimal table earlier.

To use this code, just convert a decimal number digit by digit, using four bits for each digit:

$$376 = 001101110110$$

This is *not* a binary number in the traditional sense because the bits do not have the proper positional values.

BCD can be used directly—some microprocessors implement instructions for doing BCD arithmetic (or at least addition) without converting to binary.

$$0 = 0000 \quad 5 = 0101$$
$$1 = 0001 \quad 6 = 0110$$
$$2 = 0010 \quad 7 = 0111$$
$$3 = 0011 \quad 8 = 1000$$
$$4 = 0100 \quad 9 = 1001$$

FIGURE 2.5-1:  BCD code.

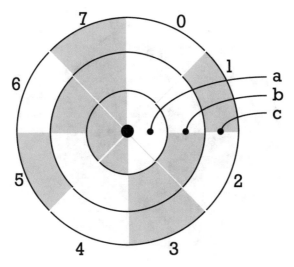

FIGURE 2.5-2: Shaft encoder (binary).

Gray codes are also called *unit-distance* codes because only one bit changes as you step from one code element to another. This is best understood by looking at a shaft encoder.

Figure 2.5-2 illustrates a shaft encoder operating in straight binary. The shaded areas are electrically conductive. The heavy dots touch these conductive and non-conductive areas and "read" the bits represented. Hence, if the contacts are touching the encoder in the middle of the "1" sector, they "see" the value 001.

But suppose the contacts are right on the boundary between the "1" sector and the "2" sector. The innermost contact clearly sees 0. But the next contact could be seeing either a 0 or a 1. Likewise, the outermost contact could be seeing either a 0 or a 1. Hence, the value being read from the disk could be 000 or 001 or 010 or 011. In other words, it's very hard to make certain where the shaft is at the moment.

But suppose we encode this shaft using a unit-distance code, a code chosen so that only one bit changes on any boundary. Figure 2.5-3 shows this.

Note that the code order is certainly not binary. But all we want is a group of three bits that tell us where the contacts are on the encoder disk. If the contacts are in the middle of the "1" sector, they see 001. If they are in the middle of the "2" section, they see 011.

Now suppose they are right on the boundary between the "1" and the "2" sectors. Only the middle contact could be in question. The innermost sees a 0 on the boundary and the outermost sees a 1. So the reading can be only 001 or 011. In other words, the sector code right on the boundary can only be 001, the code for one neighboring sector, or 011, the code for the other neighboring sector.

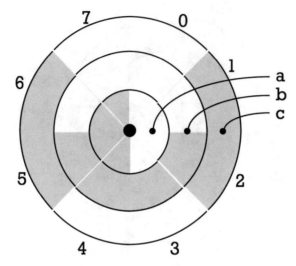

**FIGURE 2.5-3:** Shaft encoder (Gray).

There are other codes that can do all sorts of interesting things but we won't go into them here. One you might like to look up is Hamming error-correcting codes, heavily used in compact-disk recording.

## 2.6    SUMMARY

That's a lot on numbers in a small space. It's especially interesting to execute these algorithms using binary arithmetic as computers must. Complement arithmetic is particularly fascinating to study. But we've gone far enough to give you a base for the arithmetic that is needed for, say, a vending-machine controller.

We also looked at a couple of binary codes, where the bits don't have a specific relationship to numerical values. We've only just touched the area—there are many different codes out there for special purposes.

The other area that forms the background for digital logic design is Boolean algebra. We'll do a little of that in the next chapter, enough to be able to understand where some of the design techniques come from.

CHAPTER 3

# Boolean Algebra: The Formal Stuff

The mathematics of digital logic is Boolean algebra. Not that Mr. Boole set out to provide us with his algebra for this purpose, but it helps nevertheless. It took over 80 years for his work to become useful in designing switching circuits.

Keep in mind that all mathematics are "thought up." There's nothing natural about it. Calculus was invented, numbers were invented, and so on. In Boole's case, he wanted a way to formalize syllogisms such as, "All cows have four legs; this creature has only two; therefore this creature is not a cow." Exciting stuff.

We aren't going to go very far into Boolean algebra, just far enough to be dangerous. There are a few theorems that are extremely useful as we work with combinational logic. The idea of "proof" in Boolean algebra helps too. But don't worry that we'll get into a lemma–lemma–theorem–proof–remark mode and forget that our goal is to design logic circuits.

## 3.1 BOOLEAN BEGINNINGS

George Boole contrived his algebra and published it in 1854. While today we relate it mainly to binary matters such as whether the creature is or is not a cow, it doesn't have to be. We'll stick with binary.

### 3.1.1 Postulates

Any mathematics begins with an invention of some kind. Boole invented the basis for his algebra by forming a series of postulates or axioms upon which he built the rest of the algebra. One form of these contains five postulates that relate very nicely to what we'll need for developing digital logic circuits:

P1.   Variables can have only two values, defined by

$$A = 0 \; \textit{if} \quad A \neq 1,$$
$$A = 1 \; \textit{if} \quad A \neq 0.$$

P2.   The concept of a complement using the prime mark (single quote) is defined by

$$if \quad A = 0 \ then \ A' = 1,$$
$$if \quad A = 1 \ then \ A' = 0.$$

The remaining three postulates define + and • to be *or* and *and*:

P3.
$$0 • 0 = 0,$$
$$1 + 1 = 1.$$

P4.
$$1 • 1 = 1,$$
$$0 + 0 = 0.$$

P5.
$$0 • 1 = 1 • 0 = 0,$$
$$1 + 0 = 0 + 1 = 1.$$

There are many theorems that one can derive and prove from these postulates. But this isn't a math course and we are interested mainly in what Boolean algebra tells us about how to combine and simplify logic expressions.

### 3.1.2   Theorems

I'm going to list a bunch of theorems without proofs. Then I'll show you a simple way to prove a couple of them and leave the others to you.

T1.   Tells us how to simplify terms that contain a known 0 or a known 1:

$$A + 0 = A,$$
$$A • 1 = A.$$

T2.   Shows how 0 or 1 sometimes completely determines the outcome:

$$A + 1 = 1,$$
$$A • 0 = 0.$$

T3.   Allows us to simplify terms containing the same variable:

$$A + A = A,$$
$$A • A = A.$$

T4.   Complements the complement:

$$(A')' = A.$$

T5.   Combines complemented variables:

$$A + A' = 1,$$
$$A \bullet A' = 0.$$

T6.   Shows that order doesn't count (commutivity):

$$A + B = B + A,$$
$$A \bullet B = B \bullet A.$$

T7.   Shows that grouping doesn't count (associativity):

$$A + (B + C) = (A + B) + C,$$
$$A \bullet (B \bullet C) = (A \bullet B) \bullet C.$$

T8.   Is a funny form of factoring that lets us combine terms:

$$A \bullet B + A \bullet C = A \bullet (B + C),$$
$$(A + B) \bullet (A + C) = A + B \bullet C.$$

T9.   Is a powerful simplification tool because it shows how a term can cover another:

$$A + A \bullet B = A,$$
$$A \bullet (A + B) = A.$$

T10.  Is another powerful tool for combining terms:

$$A \bullet B + A \bullet B' = A,$$
$$(A + B) \bullet (A + B') = A.$$

T11.  Is DeMorgan's Theorem and tells how to complement a whole expression:

$$(A + B + C + \cdots)' = A' \bullet B' \bullet C' \bullet \cdots,$$
$$(A \bullet B \bullet C \bullet \cdots)' = A' + B' + C' + \cdots.$$

T12.  Is a generalized form of DeMorgan's Theorem that is easier to say in words than in mathematical form. To complement a complete expression,

   (a) put in all parentheses to group terms correctly—don't rely on the assumed hierarchy of $\bullet$ over $+$,

   (b) change all the $\bullet$ to $+$ and all the $+$ to $\bullet$, and

   (c) change all complemented variables to uncomplemented ones and vice versa.

   (d) Now clean up the parentheses if you need to.

Whew! That's enough of that! But now that we have this Boolean algebra laid out (rather sloppily, but in a useful way), how do we use it? In the next section, I'll demonstrate a simple method of proof, show an example of how these theorems help reduce the complexity of a logic circuit, and do one complementation using the generalized DeMorgan's Theorem.

## 3.2  USING BOOLE

Let's use Boolean algebra to prove one of the theorems. Not that I don't believe the mathematics, but I do want to demonstrate a way of showing that something is true (or false, for that matter).

The simplest approach, which is almost a no-brainer, uses *perfect induction* to demonstrate that every possible combination of 0s and 1s yields the correct result. Perfect induction is most easily carried out by writing a *truth table* that tells the whole truth and nothing but the truth.

### 3.2.1  A Proof

I'll prove using perfect induction the second half of theorem T8, which says $(A + B) \bullet (A + C) = A + B \bullet C$. Figure 3.1-1 shows my truth table for this. Here's how I developed the table:

- Create the "input" columns A, B, and C by listing all eight of the possible arrangements of 0s and 1s. (There are three variables, each of which can have one of two values, so there are $2^3$ rows in this truth table.) I list them in ascending binary number order.

| A | B | C | X=<br>A+B | Y=<br>A+C | X•Y | A | B•C | Z=<br>A+Z |
|---|---|---|-----|-----|-----|---|-----|-----|
| 0 | 0 | 0 | 0 | 0 | 0 | 0 | 0 | 0 |
| 0 | 0 | 1 | 0 | 1 | 0 | 0 | 0 | 0 |
| 0 | 1 | 0 | 1 | 0 | 0 | 0 | 0 | 0 |
| 0 | 1 | 1 | 1 | 1 | 1 | 0 | 1 | 1 |
| 1 | 0 | 0 | 1 | 1 | 1 | 1 | 0 | 1 |
| 1 | 0 | 1 | 1 | 1 | 1 | 1 | 0 | 1 |
| 1 | 1 | 0 | 1 | 1 | 1 | 1 | 0 | 1 |
| 1 | 1 | 1 | 1 | 1 | 1 | 1 | 1 | 1 |

$$\uparrow \underline{\qquad} = \underline{\qquad} \uparrow$$

FIGURE 3.1-1:  Proof of T8 (2nd part).

- Compute two columns of the parenthesized combinations $(A + B)$ and $(A + C)$ using postulates P3, P4, and P5 that tell how the + operation works.
- Now combine these two columns using the • operation and postulates P3, P4, and P5.

That's the result for the left-hand side of the = sign of the theorem I'm proving. Now do the same for the right-hand side:

- Repeat the A column for convenience.
- Compute the value of B•C using postulates P3, P4, and P5.
- Now combine these two columns using the + operation and postulates P3, P4, and P5.

That's the result for the right-hand side of the = sign. Are the two columns equal all the way down? If so, we have proven the theorem by showing that it works for every possible combination of values of A, B, and C. That's proof by perfect induction.

Now let's put all this to better use.

### 3.2.2   Logic Simplification

A certain logic dolt has created the circuit shown in Fig. 3.1-2 to implement the function $Z = A•B•C' + A•C + A'•C$. Notice that this circuit requires four gates and two inverters.

I'll use Boolean algebra to reduce the number of gates if possible. The process is by no means *algorithmic*, which means I can't write any definitive rules to tell you how to proceed. This depends on a certain amount of *Aha!* as you see what terms combine and simplify.

Our goal is to reduce the number of gates, which means combining terms to get rid of as many + and • operations as we can. Here's the original expression:

$$Z = A \bullet B \bullet C' + A \bullet C + A' \bullet C.$$

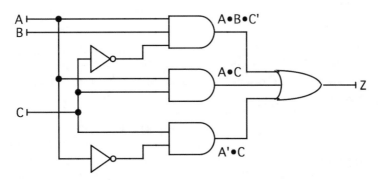

FIGURE 3.1-2: A logic circuit.

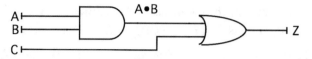

FIGURE 3.1-3: A logic circuit simplified.

Apply T10 to the last two terms, reducing them to simply C:

$$Z = A \bullet B \bullet C' + C.$$

Use T7 and T6 to regroup the remaining terms:

$$Z = C + C' \bullet (A \bullet B).$$

Now *expand* this expression somewhat by using T8 to get C and C' together:

$$Z = (C + C') \bullet (C + A \bullet B).$$

But T5 tells us that $C + C' = 1$:

$$Z = 1 \bullet (C + A \bullet B).$$

Finally, T1 will get rid of the 1 and I'll remove the parentheses too. The result is a much simpler expression, or at least it looks that way:

$$Z = C + A \bullet B.$$

The real test, though, is whether the gate circuit is simpler. Look at Fig. 3.1-3; did I succeed? Reducing the circuit from six logic gates to only two isn't too bad.

Are we going to have to do this stuff every time we want to look for a simpler logic circuit? Fortunately not. There are several neat algorithms that will do the job for us. One of them uses our ability to perceive regular combinations of things graphically; the other uses a computer tabulation.

### 3.2.3   Applying DeMorgan

DeMorgan's Theorem is quite useful because complementing an expression arises from time to time. To illustrate the use of the generalized form of the theorem, I'll do an example.

Consider the logic expression

$$Z = A + B \bullet C' + (D' + E \bullet F' \bullet G)'$$

that I want to complement:

$$Z' = [A + B \bullet C' + (D' + E \bullet F' \bullet G)']'.$$

First, I will install the parentheses to group terms so that I am not relying on the assumed hierarchy of ● over +:

$$Z' = [A + (B \bullet C') + (D' + (E \bullet F' \bullet G))']'.$$

Then I will unprime the primed terms and prime the unprimed ones and also exchange ● and + operators. Note that there is a "big" term in the parentheses at the right that has a prime on it (just after double parentheses). I remove this prime and leave all the terms and operators inside alone.

$$Z' = A' \bullet (B' + C) \bullet (D' + (E \bullet F' \bullet G)).$$

Finally, I clean up any unnecessary parentheses:

$$Z' = A' \bullet (B' + C) \bullet (D' + E \bullet F' \bullet G).$$

The result is now the complement of the original expression. This can be carried to any depth of the parentheses or complication of expression, but "real-world" logic functions rarely get that messy.

I suppose you are ready at this point to get on with the process of finding better ways of reducing and simplifying logic circuits? OK, but there are some things that we need to define before we can do that.

## 3.3   CANONICAL FORMS

The starting point for most of the algorithms for simplification is an expression in *canonical form*. While that sounds like something awful, it isn't. It's just a neat starting point that gets the algorithmic process off to a proper start.

A one-n canon is a rule or law, often applied to church law. A two-n cannon goes BANG. To get through this, we have some new terms to learn. In every field there are terms that are specific to that field, that help define things sharply. Digital logic is no different.

### 3.3.1   Definitions

Here are the important definitions that we'll need:

*Product term* is a term where the variables are connected by *and* operators, terms such as A●B, C'●D●E, and even the simple term F.

*Sum term* is a term where the variables are connected by *or* operators, terms such as A + B, C' + D + E, and even the simple term F.

*Literal* is any appearance of a variable in a term, whether primed or not. In the terms above, the literals are A, B, C′, D, E, and F. If the variable appears both primed and unprimed, as in G and G', each is a literal.

*Sum-of-products* is an expression consisting of product terms linked by *or* operators, as in A•B + C′ • D • E + F.

*Product-of-sums* is an expression consisting of sum terms linked by *and* operators, as in (A + B) • (C′ + D + E)•F.

*Normal term* is a term, either a product term or a sum term, in which no variable appears more than once. For example, the term A•B•C•C is not normal because we can use theorem T3 to remove one appearance of C to yield A•B•C. Also, the term D•E•E′ is not normal because we can use theorem T5 to reduce the term to 0. Non-normal terms are pretty obvious and represent unnecessary complications in our logic circuits.

*Minterm* is a normal product term that contains all the variables of the problem. For example, if the problem contains the variables A, B, and C, then all minterms must contain these three variables, with or without primes. So A′•B•C and A•B′•C′ are two possible minterms in this case. The term A•B is not a minterm in this case because it does not contain C. To make life a little more orderly, we usually write the variable names in alphabetical order.

*Maxterm* is a normal sum term that contains all the variables of the problem. For example, if the problem contains the variables A, B, and C, then all maxterms must contain these three variables, with or without primes. So A′ + B + C and A + B′ + C′ are two possible maxterms in this case. The term A + B is not a maxterm in this case because it does not contain C. To make life a little more orderly, we usually write the variable names in alphabetical order.

*Minterm number* is a way of coding the minterm to make the notation simpler. (This is very useful when applying a simplification algorithm.) Suppose the problem contains three variables, A, B, and C. We must write these in the same order in every minterm. When we do this, we can give the minterm a numerical code. For example, the minterm A′•B•C receives a code value of 3. This is derived by reading variables with primes as 0 and variables without primes as 1. Hence, A′•B•C yields 011, which in decimal is 3. To indicate this code, we use the summation symbol $\sum$. Hence A′•B•C = $\Sigma(3)$.

*Maxterm number* is a way of coding the maxterm to make the notation simpler. Again suppose the problem contains three variables, A, B, and C. We must write these in the same order in every maxterm. Then we encode the maxterm by reading primed variables as 1 and unprimed variables as 0. Note that this is opposite to the encoding of minterms! Hence, the maxterm A′ + B + C receives a code value of 4 because the encoding yields

| A B C | Minterm | Minterm number | Maxterm | Maxterm number |
|-------|---------|----------------|---------|----------------|
| 0 0 0 | A'•B'•C' | Σ(0) | A+B+C | Π(0) |
| 0 0 1 | A'•B'•C | Σ(1) | A+B+C' | Π(1) |
| 0 1 0 | A'•B•C' | Σ(2) | A+B'+C | Π(2) |
| 0 1 1 | A'•B•C | Σ(3) | A+B'+C' | Π(3) |
| 1 0 0 | A•B'•C' | Σ(4) | A'+B+C | Π(4) |
| 1 0 1 | A•B'•C | Σ(5) | A'+B+C' | Π(5) |
| 1 1 0 | A•B•C' | Σ(6) | A'+B'+C | Π(6) |
| 1 1 1 | A•B•C | Σ(7) | A'+B'+C' | Π(7) |

FIGURE 3.2-1: Minterms & maxterms (3 variables).

100, which in decimal is 4. To indicate this code, we use the product symbol Π. Hence $A' + B + C = \Pi(4)$.

Figure 3.2-1 shows all the minterms and maxterms of three variables, along with their number codes.

This is a good time to make use of some of these new terms.

### 3.3.2   Example of Minterms and Maxterms

Suppose that we have gotten from somewhere a logic expression that we want to represent in canonical form. Here's an example:

$$Z = A \bullet B \bullet C' + A \bullet C + A' \bullet C.$$

(You might recognize it as the expression we simplified earlier.)

I will expand it into minterm form by expanding all terms so they all have all three variables of the problem, namely, A, B, and C.

The term A•C is missing B. By theorem T1 I can expand this to A•1•C. Then I can use theorem T5 to write 1 in terms of B as $B + B'$. Now the term has become A•(B + B')•C. Theorem T8 lets me expand this into two terms A•B•C + A•B'•C.

I'll do the same thing with the term A'•C, expanding it into A'•B•C + A'•B'•C.

The final result of this expansion is

$$Z = A \bullet B \bullet C' + A \bullet B \bullet C + A \bullet B' \bullet C+$$
$$A' \bullet B \bullet C + A' \bullet B' \bullet C.$$

(If we happen to create two identical terms in this process, which often happens, theorem T3 allows us to eliminate all but one of them.)

Now we can encode these terms using minterm numbers:

$$Z = \sum (6) + \sum (7) + \sum (5) + \sum (3) + \sum (1),$$

which we reorganize as

$$Z = \sum (1, 3, 5, 6, 7).$$

But this notation has created a problem: we can't tell what the original letters were. So we add to the left-hand side of the equation a function-like notation:

$$Z(A, B, C) = \sum (1, 3, 5, 6, 7)$$

This makes it clear what the codes mean. Notice that the order of the variable names is very important.

Here's an example of maxterm coding:

$$X = (G + H) \bullet (G' + H + J) \bullet (H + J),$$
$$= (G + H + J \bullet J') \bullet (G' + H + J) \bullet (G \bullet G' + H + J),$$
$$= (G + H + J) \bullet (G + H + J') \bullet (G' + H + J) \bullet$$
$$(G + H + J) \bullet (G' + H + J),$$
$$= (G + H + J) \bullet (G + H + J') \bullet (G' + H + J),$$
$$= \prod (0) \bullet \prod (1) \bullet \prod (4),$$
$$X(G, H, J) = \prod (0, 1, 4).$$

Notice that I dropped two terms that were repeated.

### 3.3.3   Complements and Conversions

The complement of a logic function in its encoded form is very easy to do. To complement a function in either $\Sigma$ or $\Pi$ form, replace the numerical codes in the list with all the codes that aren't there.

OK, which ones aren't there? Suppose we have three variables. Then the possible codes are the integers 0 through 7. Four variables would require codes from 0 to 15 (because $2^4 = 16$), and so on.

Hence, the complement of $Z(A,B,C) = \Sigma (1,3,5,6,7)$ is

$$Z'(A, B, C) = \sum (0, 2, 4)$$
$$= A' \bullet B' \bullet C' + A' \bullet B \bullet C' + A \bullet B' \bullet C'$$

and the complement of $X(G,H,J) = \Pi\,(0,1,4)$ is

$$X'(G, H, J) = \prod(2, 3, 5, 6, 7)$$
$$= (G + H' + J) \bullet (G + H' + J') \bullet (G' + H + J') \bullet$$
$$(G' + H' + J) \bullet (G' + H' + J').$$

Proving these can be done by writing truth tables.

Conversion from the product form to the sum form is also straightforward. Change the product or sum symbol to the other symbol and replace the numerical codes with all the codes that aren't there. For example,

$$Z(A, B, C) = \sum(1, 3, 5, 6, 7)$$
$$= \prod(0, 2, 4).$$

Does all this come together? Yes! We often have as our starting point a truth table that describes the logic function we want. Our goal is to develop a logic circuit in hardware. So we need to look at truth tables.

## 3.4   TRUTH TABLES

A truth table is a tabulation of *all* possible outcomes for a particular logic operation. This tabulation leads directly to the canonical forms of the logic function, either in minterm or maxterm form or in those forms encoded numerically.

The truth table in Fig. 3.3-1 represents a logic function W that I would like to implement. I'll demonstrate the various canonical forms by reading them directly from the table.

I can choose to implement this function either through the minterm or the maxterm form. Let's do the minterm form.

First, I will read from the truth table the numerical codes for the 1s of W. I do this by reading the binary number in the input columns. For example, the first 1 in the W column is four rows from the top. The bits of the input are 0011, which in decimal is 3. The minterm canonical form is

$$W(P, Q, R, S) = \sum(3, 5, 7, 8, 14, 15).$$

If I want to expand this into literal form, I read each numerical code. The 3 code becomes P'•Q'•R•S, for example. This effort gives

$$W = P' \bullet Q' \bullet R \bullet S + P' \bullet Q \bullet R' \bullet S + P' \bullet Q \bullet R \bullet S$$
$$+ P \bullet Q' \bullet R' \bullet S' + P \bullet Q \bullet R \bullet S' + P \bullet Q \bullet R \bullet S.$$

If I want the maxterm form, I can convert the minterm form as we did in the preceding section.

| P | Q | R | S | W |
|---|---|---|---|---|
| 0 | 0 | 0 | 0 | 0 |
| 0 | 0 | 0 | 1 | 0 |
| 0 | 0 | 1 | 0 | 0 |
| 0 | 0 | 1 | 1 | 1 |
| 0 | 1 | 0 | 0 | 0 |
| 0 | 1 | 0 | 1 | 1 |
| 0 | 1 | 1 | 0 | 0 |
| 0 | 1 | 1 | 1 | 1 |
| 1 | 0 | 0 | 0 | 1 |
| 1 | 0 | 0 | 1 | 0 |
| 1 | 0 | 1 | 0 | 0 |
| 1 | 0 | 1 | 1 | 0 |
| 1 | 1 | 0 | 0 | 0 |
| 1 | 1 | 0 | 1 | 0 |
| 1 | 1 | 1 | 0 | 1 |
| 1 | 1 | 1 | 1 | 1 |

FIGURE 3.3-1: A logic function.

Notice that you can read the minterm form (unencoded) directly from the truth table in Fig. 3.3-1. For example, the first 1 in the W column is on the 0011 row. This row represents P'•Q'•R•S by reading the 0s as primed variables and the 1s as unprimed variables.

We will use the canonical forms, both encoded and unencoded, as the starting point for the simplification algorithms in the next chapter.

## 3.5   SUMMARY

Boolean algebra can be very extensive. Our interest in it is for simplifying the logic functions, with a goal of producing simpler logic circuits. While "simpler" probably means "less costly," Boolean algebra doesn't guarantee that.

Most of our simplification effort will be via algorithms. These all start with one of the canonical forms ("conforming to law"), either in literal or numerically-encoded form. Both minterm and maxterm forms are useful, but I'll tend to emphasize the minterms (for no particular reason other than bias).

Now that we have the basics—binary numbers, arithmetic, Boolean algebra, simplification theorems, and canonical forms—it's time to put these to orderly use designing combinational circuits.

CHAPTER 4

# Combinational Logic: No Time Like the Present

I suppose we've had enough of an introduction to various basics such as numbers and Boolean algebra. You are ready, I'm sure, to do something more interesting. Combinational logic is more interesting. Why? Because I said so and because you are now four chapters into this book.

What is combinational logic? It's logic that combines inputs to get outputs. Sort of obvious, eh? But more importantly, it does not involve time or memory. In other words, the past does not influence the output. Only the present moment counts. The outputs are directly and 100% dependent on the inputs right now.

That's not to say that combinational logic functions in zero time. The outputs do not appear at the other end of the circuit instantaneously. Changes take time to pass through the circuit. But this is not the "time or memory" that I have just mentioned. This is simply delay that is inherent in all logic devices.

Just for the record, logic circuits that involve time and memory are called *sequential logic*. We'll go into those starting in Chapter 6.

## 4.1 GATES AND SYMBOLS

All of our elementary combinational logic will be made up of several types of gates. The basic ones are And, Or, and Not. To these we add Exclusive Or, although it is much less commonly used. I will also introduce the Nand and Nor gates separately because they are used in a different manner.

### 4.1.1 Basic Gates

Figure 4.1-1 shows the four basic gates and their truth tables. Here's a brief comment about each:

*And* requires all inputs to be 1 for the output to be 1. Gates are available with a number of different input configurations, including two, three, four, and eight inputs.

And

| a | b | c |
|---|---|---|
| 0 | 0 | 0 |
| 0 | 1 | 0 |
| 1 | 0 | 0 |
| 1 | 1 | 1 |

Or

| a | b | c |
|---|---|---|
| 0 | 0 | 0 |
| 0 | 1 | 1 |
| 1 | 0 | 1 |
| 1 | 1 | 1 |

Not

| a | c |
|---|---|
| 0 | 1 |
| 1 | 0 |

Exclusive Or

| a | b | c |
|---|---|---|
| 0 | 0 | 0 |
| 0 | 1 | 1 |
| 1 | 0 | 1 |
| 1 | 1 | 0 |

FIGURE 4.1-1: Basic gates.

*Or* requires only one input to be a 1 for the output to be 1. To say this another way, only if all inputs are 0 will the output be 0. Like And gates, Or gates are available in several different configurations.

*Not* merely inverts the input to produce the output. Generally, though, a Not also serves as a *buffer* to provide a stronger signal that can connect to inputs of more gates or to devices such as LEDs. It's nice to avoid the Not where possible because it takes up space on the circuit board.

*Exclusive Or* will produce a 1 if the two inputs are different. Its name comes from the fact that its truth table is the same as for the Or gate excluding the "both" case. The only common configuration of this gate is with two inputs.

While we can build everything we need with these four (actually, even without the Exclusive Or), we don't. Let's see what Nand and Nor gates offer.

## 4.1.2   Nand and Nor

Two other gates are important for a number of reasons. They are the most common gates in use, but I have separated them from the first group because the differences are very important.

| a | b | c |
|---|---|---|
| 0 | 0 | 1 |
| 0 | 1 | 1 |
| 1 | 0 | 1 |
| 1 | 1 | 0 |

| a | b | c |
|---|---|---|
| 0 | 0 | 1 |
| 0 | 1 | 0 |
| 1 | 0 | 0 |
| 1 | 1 | 0 |

FIGURE 4.1-2: Nand & Nor.

Nand and Nor gates are a little cheaper than the gates we've just looked at because they are slightly simpler inside. This simplicity also makes them a little faster than And and Or gates. In fact, And and Or gates are have the same internal electronics as Nand and Nor, but they have an inverter added to their outputs.

Figure 4.1-2 shows the two gates. Here's how they function:

*Nand* is the same as And but with its output inverted. Notice that the output will be 0 only if both inputs are 1.

*Nor* is the same as Or but with its output inverted. The output is 1 only if both inputs are 0.

If you were reading carefully a couple of paragraphs ago, you might have caught what now seems like an error. I said that the And gate was a Nand gate with the addition of an output inverter.

Whoa, hold on, it sure looks in Fig. 4.1-2 like there is a bubble on the output of the Nand gate. Isn't this an inverter? Well, yes, but it is a *logical* inverter, In other words, it shows what is done logically to the And gate to get a Nand gate. Electrically, the internal hardware of an And gate automatically inverts, requiring the addition of an inverter to get the correct output. So the Nand gate naturally inverts, while the And gate inverts and then reinverts. That's why the And gate is the slower of the two.

Another interesting feature of the Nand gate is that it is a "universal" gate. This says that every possible logic function can be implemented using only Nand gates. No inverters are required. This cannot be said about And gates. Implementation of any logic function using And gates may also require Or and Not gates.

Similarly, the Nor gate is also universal, so anything built solely with Nand gates can also be done with only Nor gates.

There is another way to look at these two gates. Consider the Nand gate shown in Fig. 4.1-3. Next to it I have drawn an Or gate with bubbles on its input. The truth table below has

| a | b | c |   | a | b | a' | b' | c |
|---|---|---|---|---|---|----|----|---|
| 0 | 0 | 1 |   | 0 | 0 | 1  | 1  | 1 |
| 0 | 1 | 1 |   | 0 | 1 | 1  | 0  | 1 |
| 1 | 0 | 1 |   | 1 | 0 | 0  | 1  | 1 |
| 1 | 1 | 0 |   | 1 | 1 | 0  | 0  | 0 |

FIGURE 4.1-3: Two Nand gates.

a couple of extra columns. The inputs are still $a$ and $b$ but these inputs are complemented on their way into the Or gate. These internal inputs are shown in the middle columns. The output $c$ is the Or of these internal inputs. Notice that the output of this inverted-input Or gate is identical to that of the Nand gate.

Consider the Nor gate shown in Fig. 4.1-4. I've done the same thing. Note that the output of this inverted-input And gate is identical to that of the Nor gate.

What this says is that we can convert between the two forms when we need to. The procedure is simple: move the bubble from the output to *all* the inputs and change the basic gate from And to Or or vice versa.

This provides a simple way of converting the form of a logic circuit. Look at Fig. 4.1-5. The first circuit is made entirely of Nand gates. In the second form of the circuit I have moved the output bubble through the right-hand gate to its inputs and then changed its symbol from an And to an Or. This gate is still logically a Nand gate, though.

In the third circuit I have let the bubbles "eat up one another" so they vanish. After all, a signal that is inverted on the way out of one gate and inverted again on the way into another

| a | b | c |   | a | b | a' | b' | c |
|---|---|---|---|---|---|----|----|---|
| 0 | 0 | 1 |   | 0 | 0 | 1  | 1  | 1 |
| 0 | 1 | 0 |   | 0 | 1 | 1  | 0  | 0 |
| 1 | 0 | 0 |   | 1 | 0 | 0  | 1  | 0 |
| 1 | 1 | 0 |   | 1 | 1 | 0  | 0  | 0 |

FIGURE 4.1-4: Two Nor gates.

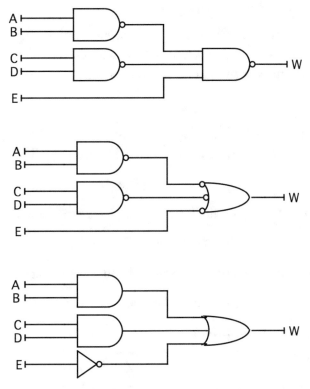

FIGURE 4.1-5: Converting to And–Or.

hasn't really been changed. Notice the handling of the input E, which now needs a Not gate to account for the inversion.

Violá! My circuit is now in And–Or form. So, you say? Well, the result of the logic minimization procedures is generally And–Or or Or–And logic. By reversing what we have just done, we have a way to convert to Nand–Nand or Nor–Nor logic. We'll do this later when we start to minimize gate circuits.

## 4.1.3   Levels of Logic

Combinational circuits don't involve time or memory as I've said. But there is still a time delay through the electronics of the gates. It is fairly common to keep these delays to a minimum by avoiding signal paths that pass through more than two gates. This is called "two-level" logic.

Look at the top circuit of Fig. 4.1-5 again. Note that the signals A, B, C, and D all pass through two gates on their way to influencing the output W. While E passes through only one gate, the longest path governs. Hence, the circuit of Fig. 4.1-5 is two-level logic.

Now let's take some circuits and analyze them.

## 4.2    LOGIC CIRCUIT ANALYSIS

Analysis is dull! But it's necessary for a couple of reasons. First, it helps us learn to deal with the symbols and the signals of digital logic. Second, it is necessary during design because the designer should analyze the design to make sure it does what is wanted.

### 4.2.1   Analysis Example I

Analyze the circuit of Fig. 4.2-1(a). One way to keep up with what is going on is to write intermediate results on the drawing. I've done this in Fig. 4.2-1(b). The result is

$$Z = A \bullet B \bullet (B' + C').$$

This can be "improved" by expanding it ("multiplying out" using theorem T8):

$$Z = A \bullet B \bullet B' + A \bullet B \bullet C'.$$

The first term is zero by theorem T5 so the circuit is providing the output

$$Z = A \bullet B \bullet C'.$$

Let's express this in canonical form:

$$Z(A, B, C) = \sum(6).$$

Can you provide this output $Z$ more simply than my circuit does?

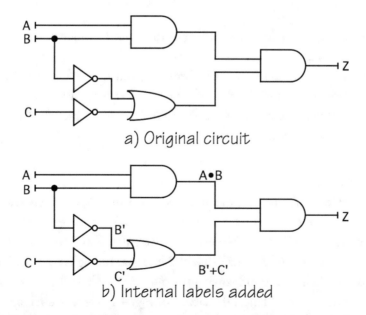

a) Original circuit

b) Internal labels added

FIGURE 4.2-1:  Analysis example I.

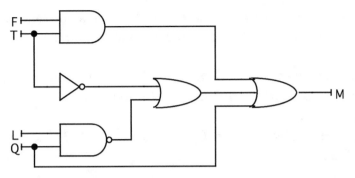

FIGURE 4.2-2: Analysis example II.

## 4.2.2    Analysis Example II

The circuit of Fig. 4.2-2 is three-level logic that has a feature that you should be suspicious of. Notice that the input Q enters the Nand gate and also goes around that gate and enters the final Or gate. This is in general bad news in a logic circuit. Not only does it indicate excessively complicated logic, it also sets up possible time-delay problems called *hazards* that can cause errors in circuits that the output is driving.

But even though the circuit is poorly designed, we can still analyze it.

- The output of the And gate at the top is F•T.
- The output is the Nand gate is $(L•Q)' = L' + Q'$.
- The inner Or gate produces $T' + (L' + Q')$.
- Hence the output $M = F•T + T' + L' + Q'$.

This result can be expressed in canonical form, but it takes some doing. We could expand each of the terms so that all the letters are present. That takes some writing; let's do it with the term F•T:

- I need to include the variables L and Q, so I'll use theorem T5 to do this: $F•T = F•(L + L')•(Q + Q')•T$. Each of those added terms is equal to 1 so this doesn't alter the original term at all.
- Now expand these by "multiplying" using theorem T8 to yield the minterms $F•L'•Q'•T + F•L'•Q•T + F•L•Q'•T + F•L•Q•T$. Notice that I have kept the letters in alphabetical order so that I can use the numerical coding correctly.
- Read the terms in binary (primes are 0), which gives the codes $\Sigma(9,11,13,15)$.

But there is an easier way. Encode the term to be expanded, keeping the letters in order and leaving blanks where letters are missing. Doing that to the F•T term gives F__T. Convert this to the binary notation for the minterm code: 1__1. Now mentally fill in the missing bits in all possible combinations: 1001, 1011, 1101, and 1111, which convert to 9, 11, 13, and 15.

Now I'll do this with the term T'. First, this is ___T'. Changing to binary gives ___0. The eight possible codes are 0000, 0010, 0100, 0110, 1000, 1010, 1101, and 1110, which are 0, 2, 4, 6, 8, 10, 12, and 14.

Similarly, the term Q' converts into 0, 1, 4, 5, 8, 9, 12, and 13. The term L' converts into 0, 1, 2, 3, 8, 9, 10, and 11.

Combining all these, putting them in order, and dropping duplicate, the final result is

$$M(F, L, Q, T) = \sum (0, 1, 2, 3, 4, 5, 6, 8, 9, 10, 11, 12, 13, 14, 15).$$

This function can be implemented a lot more simply too.

### 4.2.3   Analysis Example III

Figure 4.2-3 is Nand–Nand logic. Find the minterm canonical form in both literal and coded forms:

- Convert the output gate to the other form of the Nand by moving the bubble through to the inputs and changing the basic gate to Or.

- Remove pairs of bubbles since each Nand gate inverts its output and the output gate inverts its inputs.

- The outputs of the three And gates (after bubble conversion) are $W' \bullet X$, $X' \bullet Y' \bullet Z$, and $W \bullet X' \bullet Y$.

- The output $P = W' \bullet X + X' \bullet Y' \bullet Z + W \bullet X' \bullet Y$.

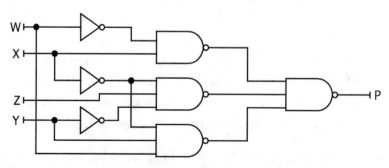

FIGURE 4.2-3:  Analysis example III.

Each term now needs to be expanded to include all the variables W, X, Y, and Z in that order:

$$W' \bullet X = W' \bullet X \bullet (Y + Y') \bullet (Z + Z')$$
$$= W' \bullet X \bullet Y' \bullet Z' + W' \bullet X \bullet Y' \bullet Z +$$
$$W' \bullet X \bullet Y \bullet Z' + W' \bullet X \bullet Y \bullet Z$$
$$= \sum (4, 5, 6, 7),$$

$$X' \bullet Y' \bullet Z = (W + W') \bullet X' \bullet Y' \bullet Z$$
$$= W' \bullet X' \bullet Y' \bullet Z + W \bullet X' \bullet Y' \bullet Z$$
$$= \sum (1, 9),$$

$$W \bullet X' \bullet Y = W \bullet X' \bullet Y \bullet (Z + Z')$$
$$= W \bullet X' \bullet Y \bullet Z' + W \bullet X' \bullet Y \bullet Z$$
$$= \sum (10, 11).$$

Combining these results into both minterm canonical form and its coding gives

$$P(W, X, Y, Z) = W' \bullet X' \bullet Y' \bullet Z + W' \bullet X \bullet Y' \bullet Z' +$$
$$W' \bullet X \bullet Y' \bullet Z + W' \bullet X \bullet Y \bullet Z' +$$
$$W' \bullet X \bullet Y \bullet Z + W \bullet X' \bullet Y' \bullet Z +$$
$$W \bullet X' \bullet Y \bullet Z' + W \bullet X' \bullet Y \bullet Z$$
$$= \sum (1, 4, 5, 6, 7, 9, 10, 11).$$

That's enough analysis to let you see what needs to be done. I've also included the "bubble" conversion of the Nand gate, which sometimes simplifies analysis. And I've expanded results into canonical form in a couple of different ways. It's time to design a few circuits.

## 4.3   MINIMIZATION

Whenever we design a logic circuit, we need to wonder whether we have chosen the design to make the circuit as simple as possible. After all, there is little reason for making a circuit unnecessarily complex. It adds to the cost of the circuit, it increases the chances of failure, and it makes it hard to maintain.

But what do we mean when we say that we want a design to be as simple as possible? What does "simple" mean? There are lots of criteria, some of which are fairly easy to achieve. Here are some:

*Minimum cost.* This one seems pretty clear, but it is fairly hard to achieve. For example, a circuit might cost less, even though it is logically more complex, because it makes use of a cheaper but more versatile logic element. Or a circuit may be more costly only because it takes more space on the printed-circuit board. We don't have an algorithm for minimizing cost except by trial-and-error.

*Minimal parts count.* It would seem that, the fewer the parts, the cheaper the circuit. But we might create excessive complexity.

*Minimal gate count.* This doesn't sound too bad. Fewer gates probably implies fewer logic elements and hence fewer packages and less board space. But not necessarily. However, reducing the gate count is algorithmic, so we can achieve this one.

*Minimal input count.* Hmmm, is this a good deal? Who cares about the number of wires entering a device? Yet a device has a fixed number of inputs. One too many would make us use either a larger device or two of the same ones. The number of inputs governs to some extent the number of traces on the printed-circuit board too. Traces take space, and space costs. Reducing the input count is algorithmic too, so this one is possible.

The popular minimization techniques reduce to some minimal value both the gate count and the input count. We will assume that this will also minimize the cost, but not necessarily.

I'm going to explain just one minimization technique, the map method. I'll mention another at the end of this section.

### 4.3.1   Map Minimization—Two Variables

The map method of minimization makes use of our ability to perceive patterns. The method starts with a logic function in either truth table or canonical form. The result is either And–Or or Or–And two-level logic, which can be converted readily into Nand–Nand or Nor–Nor form.

Let's start with a simple Venn diagram, as shown in Fig. 4.3-1. There are two circular regions, A and B. If A and B are logic variables, then we consider A to be true (and hence 1) inside its circle. A is false (and hence 0) outside this circle. B similarly has two regions, 1 and 0.

I can express this information as logic functions. The region outside both circles is A′•B′, meaning not A and not B. In other words, both A and B are false in this region. Likewise, the region inside the A circle that doesn't overlap with B is A•B′. The region inside the B circle that doesn't overlap A is A′•B. Finally, the region of overlap is A•B where both are true.

Now I modify the Venn diagram by allowing the A circle to expand until it completely fills the right-hand half of the square. I also allow the B circle to expand to completely fill the bottom half of the square. The result is shown in Fig. 4.3-2. Notice that the four regions are still present, but now they are squares.

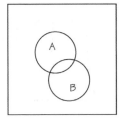

FIGURE 4.3-1: Venn diagram.

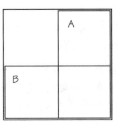

FIGURE 4.3-2: Modified Venn diagram.

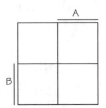

FIGURE 4.3-3: Relabeled Venn diagram.

FIGURE 4.3-4: Numbered Venn diagram = Karnaugh map.

It helps to modify the labeling a little too. Instead of labeling the rectangles inside the square, I'm going to label them outside as shown in Fig. 4.3-3. This says that the region on the right is where A = 1. Therefore, the region on the left is where A = 0. Likewise B = 1 at the bottom and B = 0 at the top.

In fact, I could write the binary values of A and B (in that order) inside the smaller squares. I've done this in Fig. 4.3-4. Then I have converted the binary numbers to decimal and written them in the corners of the squares.

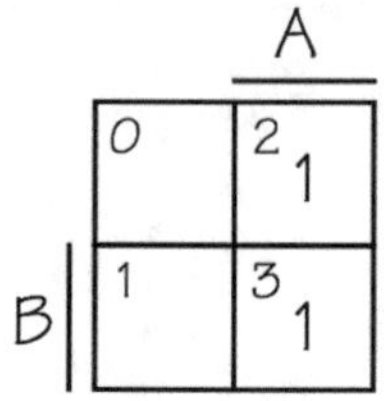

FIGURE 4.3-5: A pair.

Guess what these numbers are! Sure, the minterm codes. In the upper-right corner, for example, the code is 2 and the logic function for the square is A•B′.

Let's be more correct now and give some proper names. The small individual squares are called *cells*. The overall map is called a *Karnaugh map* after the person who first published this simplification method. (They are often called *K-maps*.)

But how do we use the K-map for simplification? I'll show two examples on this two-variable map. Suppose that the logic function Z is to be 1 when A is 1 and B is 0. That means that Z = A•B′. I'll put a 1 in the cell where A•B′ is found, namely cell 2. (See Fig. 4.3-5.)

Suppose further than Z is also to be a 1 when A and B are both 1, so Z = A•B here. I'll put a 1 in cell 3.

Now the map represents Z = A•B′ + A•B. If we apply theorem T10 to Z, we get Z = A. See if that isn't obvious from the map itself. Yup, there is a 1 everywhere in the A region, so the result is that Z = A. That is certainly simpler than the first form, which requires two And gates and an Or gate.

I argue that we could have seen this by simply looking at the map and noticing the pair of 1s that occupies the whole A region. My conclusion is that pairs of 1s convert to something simpler.

But not all pairs. Consider the map in Fig. 4.3-6 where the two 1s are on a diagonal. Here Z = AB′ + A′B. Try all the theorems you want and you won't reduce this expression. So diagonal pairs don't help.

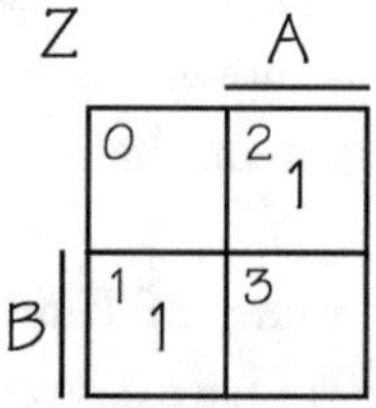

FIGURE 4.3-6: No pair.

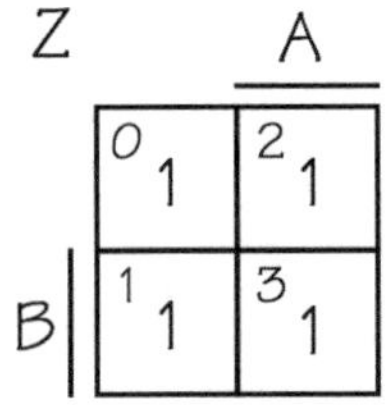

Z ⎯⎯ A

| | |
|---|---|
| 0    1 | 2    1 |
| 1    1 | 3    1 |

B

FIGURE 4.3-7:  Pair or pairs.

Here's one more two-variable map, as shown in Fig. 4.3-7. It has four 1s on it. One pair is on the right, which is A. Another pair is on the left, which is A'. Combining these gives $Z = A + A'$. Theorem T5 says this is 1, so $Z = 1$. OK, so a "pair and a pair" (pair-of-pairs) is also a simplification.

## 4.3.2  Map Minimization—Three Variables

On to a three-variable K-map. Figure 4.3-8 shows one with the minterm numbers in the cells. Since the coding of minterms depends on variable order, we must be sure to read the variables as A, B, and C in that order. To say it another way, the map is numbered in top–bottom–left order. Notice that the cells are not numbered in physical order.

Consider the map of Fig. 4.3-9. A pair is obvious, which says that Z must contain a term A•B. C is not a factor here because the two 1s are $A•B•C' + A•B•C$, which by theorem T8 can be reduced to A•B.

That takes care of two of the 1s. How about the remaining one? It won't pair with anyone because there is no 1 adjacent to it in any direction. Pairs require adjacency. So that 1 is the term A'•B'•C.

The result is that $Z = A•B + A'•B'•C$, and there is no simpler And–Or form.

Here's a trickier one, as shown in Fig. 4.3-10. The pair on the right is pretty clearly A•B'. That leaves just the 1 on the upper left. But this 1 combines with the 1 on the upper right. Oh, really? Well, the upper left cell is A'•B'•C', the upper right is A•B'•C'. Theorem T8 says that

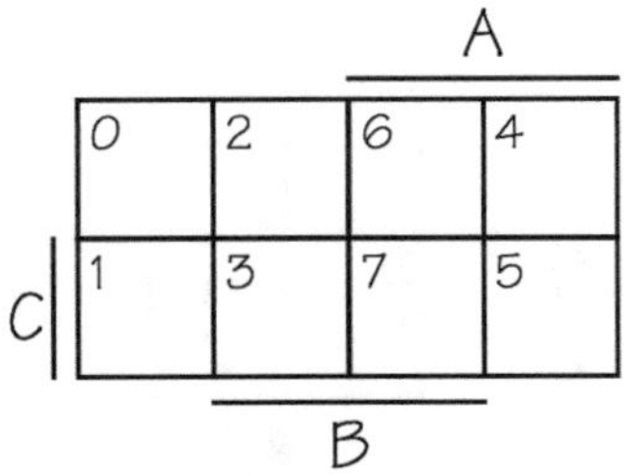

A

| 0 | 2 | 6 | 4 |
|---|---|---|---|
| 1 | 3 | 7 | 5 |

C

B

FIGURE 4.3-8:  K-Map for three variables.

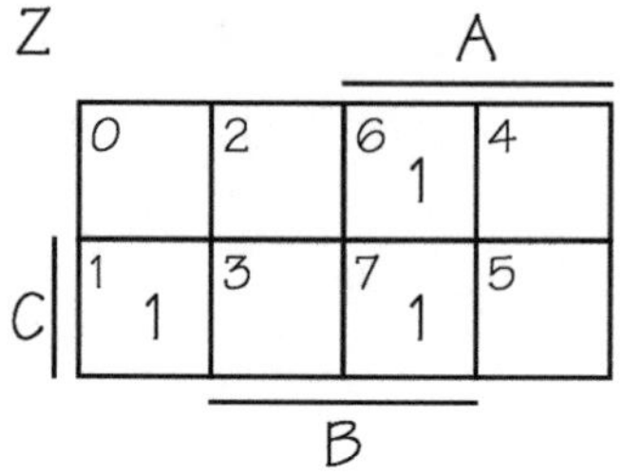

FIGURE 4.3-9: Pair plus 1.

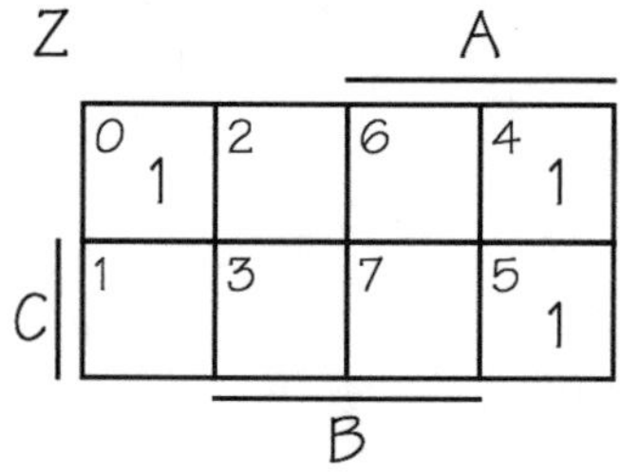

FIGURE 4.3-10: End around.

the A part will drop out, leaving just $B' \bullet C'$. In other words, pairs wrap around the back of the map as if it were drawn on a sphere.

But wait, you say! You already included the upper-right cell when you made the first pair. True, but there's no harm in Boolean algebra including a term twice. Theorem T3 says that $A \bullet B' \bullet C' + A \bullet B' \bullet C'$ is the same as $A \bullet B' \bullet C'$. So I will use a cell as often as I need to if it will simplify the result.

That makes the final result $Z = A \bullet B' + B' \bullet C'$, and no further simplification is possible, at least in And–Or form.

There's one more thing to do before going to larger maps. The map in Fig. 4.3-11 has a difference that we need to consider. Two pairs are rather evident: the vertical one on the left is $A' \bullet B'$ and the vertical one near the right is $A \bullet B$. These won't combine further.

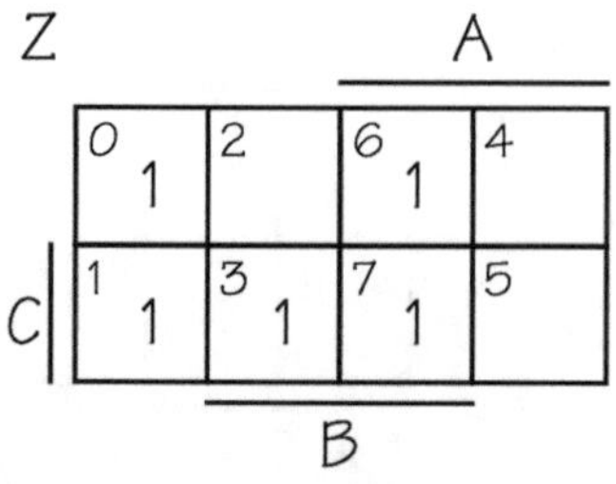

FIGURE 4.3-11: Two solutions.

What about the 1 in cell 3? It looks like it will make a pair to its left or to its right. Yup, that's true. And both logic functions will be equally simple. Both will have the same number of gates and the same number of inputs. So which do we use?

Either one will do. Sometimes we choose which one based on whether a certain input or its complement is available. But if there isn't anything to tell us which to choose, we can write both solutions:

$$Z = A' \bullet B' + A \bullet B + \begin{cases} A' \bullet C \\ B \bullet C \end{cases}$$

Don't choose both of them, though, or your result won't be minimal.

Before heading for four-variable maps, I need to show how to count inputs and gates. Look at the result for Z that we just got. This will take three And gates, one for each term. It will take one Or gate to combine the outputs of the And gates. So the gate count is four.

Inputs are counted for both the And and the Or gates. Each And gate here has two inputs for a total of six. The Or gate has an input from each And gate, three. So the input count is nine. These counts give us a way of comparing circuit minimizations.

### 4.3.3    Map Minimization—Four Variables

Now for four-variable K-maps. Figure 4.3-12 shows the general form with minterm numbers and labels on the edges. Again, the order of the variables is important. It's top–bottom–left–right here.

In the four-variable map, pairs count, of course. So do pairs-of-pairs, which are two pairs adjacent. And we'll even find pairs-of-pairs-of-pairs, which are eight cells in a rectangle. We also have two ways that edges can meet around the back of the sphere, top to bottom and left to right.

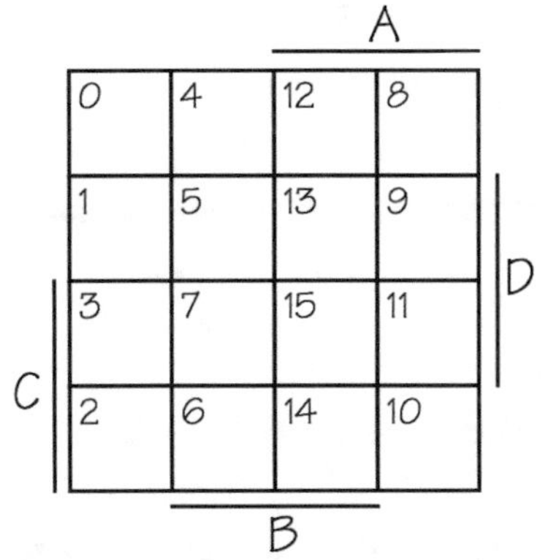

FIGURE 4.3-12: K-Map for four variables.

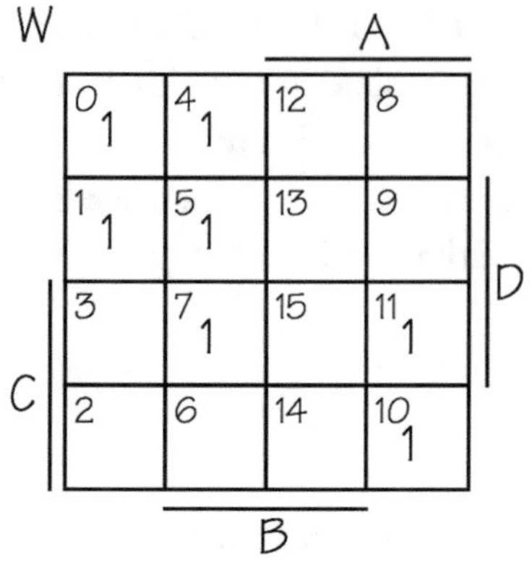

**FIGURE 4.3-13:** Pairs.

Fig. 4.3-13 maps $Z(A,B,C,D) = \Sigma\ (0,1,4,5,7,10,11)$. When maps get more complicated, it's useful to have an algorithm for finding the minimal set of terms that will include all the 1s. Here are the steps:

- Select a cell containing a 1. Find the biggest pattern of pairs that will include this cell. In other words, look for pairs, pairs-of-pairs, and so on.
- If there is just one "biggest," you've got a winner. Include that one in your result.
- If there are more than one "biggest," skip on to another cell.
- It is best to start with cells that lead to small groupings rather than more obvious large ones.

I'll go through the map of Fig. 4.3-13 as follows:

- Select cell 10. The largest pattern is the pair $A \bullet B' \bullet C$. Since that's the only "two-holer," this term ends up in the final result.
- Select cell 7. The largest pattern is the pair with the 1 in cell 5. That's the only one, so the term $A' \bullet B \bullet D$ ends up in the final result.
- Select cell 0. The largest pattern is the square, a "four-holer," so the term $A' \bullet C'$ goes into the final result.
- 

The final result is $W = A \bullet B' \bullet C + A' \bullet B \bullet D + A' \bullet C'$. This requires four gates and 11 inputs.

Now let's look at the map of Fig. 4.3-14. Remember that diagonals don't count. Here are the steps:

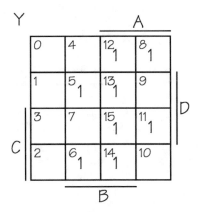

FIGURE 4.3-14:  Two squares.

- Choose cell 5. The largest pattern is the square in the middle, which is B•D.
- Choose cell 0. Hmmm, it seems to be alone. But it's not! How about wrapping around the back to cell 8? And while we are at it, how about wrapping those two cells (a pair) from top to bottom and matching up with the pair in cells 2 and 10? Right! The corners are B'•D'.

The result is $X = B \bullet D + B' \bullet D'$.

Figure 4.3-15 is a map that contains what I call a "sucker pattern." If you jump into the job ignoring the proper steps for finding pairs, you'll end up with too many terms. Recall that I suggested starting with the smaller patterns. If we do that here, things work out correctly.

- Select cell 8. The only pair is A•C'•D'.
- Select cell 5. The only pair is B•C'•D.

FIGURE 4.3-15:  Sucker map.

- Select cell 11. The only pair is A•C•D.

- Select cell 6. The only pair is B•C•D′.

- At this point, all the 1s have been included in pairs. While it is tempting to choose the vertical "four-holer," it is redundant.

The final result is $Y = A•C′•D′ + B•C′•D + A•C•D + B•C•D′$. There are lots of ways to disguise the "sucker map." And it's a good test of any computer-based minimization tool too.

Here's one more, as shown in Fig. 4.3-16:

- Select cell 0, which leads to A′•C′•D′.

- Select cell 5, which leads to A′•B•C′ and also to A′•B•D. Hmmm, which to choose? Well, the cell above 5 is already covered by the first term we found, so .... There is no way to be certain yet which of these to choose, so we'll skip it.

- Select cell 15, which leads to B•C•D.

- That leaves the 1 in cell 5 to include, and that can be done either of two ways. The minimal final result can therefore be written in two ways:

$$Z = A' \bullet C' \bullet D' + B \bullet C \bullet D + \begin{cases} A' \bullet B \bullet C' \\ A' \bullet B \bullet D \end{cases}$$

Are we going to a five-variable map now? Nope! Larger maps get pretty messy and they are beyond what I want to do in this course.

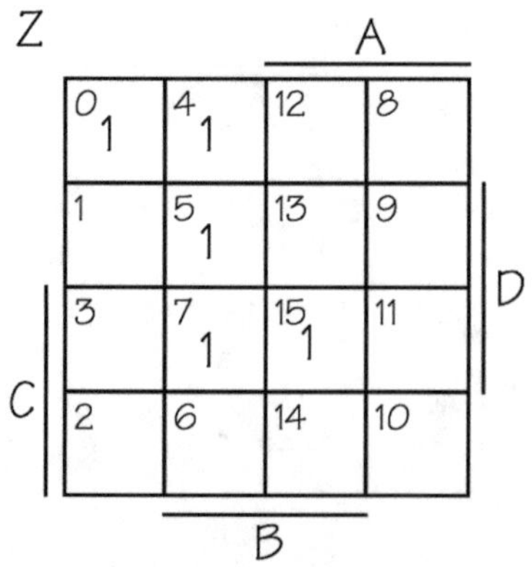

FIGURE 4.3-16: Two results.

### 4.3.4   Map Minimization—Don't-Cares

Don't-cares arise in logic design when a particular input combination can never happen or when we truly don't care what the outcome is for a particular input combination. Don't-cares are helpful because we can choose to use their cells as either 1s or 0s, depending on which will make the function minimal.

Don't-cares can be included in coded minterm notation by writing a "d" instead of a summation. They are included on K-maps by writing "x" in the appropriate cells.

Suppose we have a function described by

$$S(A, B, C, D) = \sum (4, 5, 11, 12, 13, 14) + d(7, 15).$$

I will draw the map shown in Fig. 4.3-17, where the "x" represents the two don't-care terms. Then I proceed through the map, using the don't-cares to help make pairs but never including them specifically:

- Select cell 11, which combines with one don't-care to yield the term A•C•D.

- Select cell 12, which gives a vertical four-holer and also a square four-holer at the top. Hmm, that's two "biggest," so that was not a good choice.

- Select cell 4, which leads to the square at the top, namely, B•C′.

- Select cell 14, which combines with one don't-care to give the vertical four-holer A•B.

The final result is = A•C•D + B•C′ + A•B. This uses the don't-care in cell 7 as a 0 and the don't-care in cell 15 as a 1.

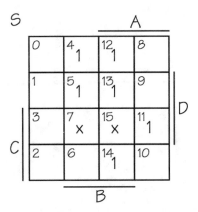

FIGURE 4.3-17: Don't-cares.

### 4.3.5    Other Minimization Methods

The map method relies on human perception. That's something that's hard to teach a computer if we want to mechanize logic circuit design. So we need something that doesn't require judgment.

One popular method is the Quine-McCluskey tabular method. This method creates tables of the minterms and then proceeds in a very regular way to combine them. It even stays out of the error that the sucker map leads to. We won't go into the tabular method here.

Way back in the beginning of this chapter I introduced the Nand and the Nor gates and said they were important. Yet all our designs so far have been And–Or designs. We need to look beyond And–Or.

## 4.4    OTHER GATE ARRANGEMENTS

And–Or is not the only way to organize two-level logic. In fact, it is one of the less popular forms. I'm going to look at two other two-level forms.

### 4.4.1    And–Or and Nand–Nand Circuits

Look back for a moment to Section 4.2.3 where we analyzed a circuit (Example III) that was in the form of Nand–Nand gating. The result of this analysis was

$$P(W, X, Y, Z) = \sum (1, 4, 5, 6, 7, 9, 10, 11).$$

The Karnaugh map for this function is shown in Fig. 4.4-1. Let's write the minimal And–Or function for this:

- Cell 10 leads to the pair W•X'•Y.

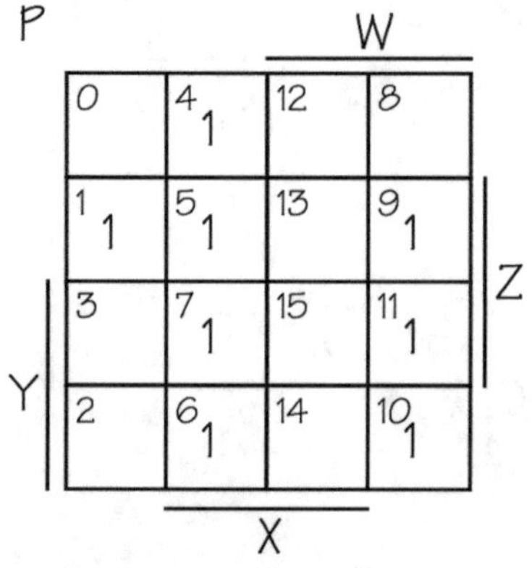

FIGURE 4.4-1: Redo example III.

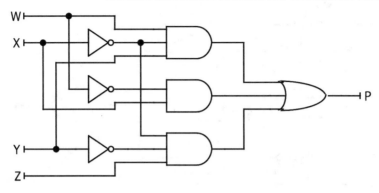

FIGURE 4.4-2: And–Or result.

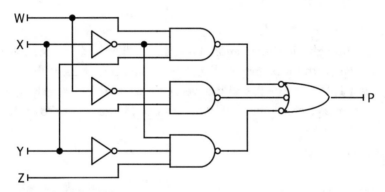

FIGURE 4.4-3: Results with double bubbles.

- Cell 4 leads to the pair-of-pairs W′•X.
- That leaves only cells 1 and 9, which form the pair X′•Y′•Z.

The result is $P = W•X′•Y + W′•X + X′•Y′•Z$. What does the circuit look like? Figure 4.4-2 is the And–Or logic that implements P.

Now consider what happens when I complement all the And outputs and also complement the inputs to the Or gate. (See Fig. 4.4-3.) Doing this changes nothing as far as the logic is concerned. But the output gate is now the other form of the Nand gate. The And gates are now also Nand gates. So the result, as shown in Fig. 4.4.4, is the Nand–Nand form that produces the function P.

What have we learned? That Nand–Nand circuits can be gotten directly from And–Or simplifications by putting bubbles in appropriate places.

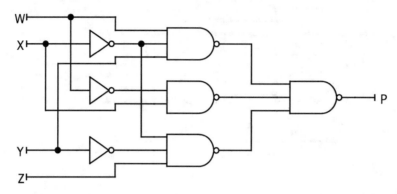

**FIGURE 4.4-4:** Result in Nand–Nor form.

## 4.4.2    Or–And and Nor–Nor Circuits

Rather than going through a long derivation, I'm going to "just do it" to develop another result of simplification. Start with the K-map of Fig. 4.4-1. Then complement the function P. That means that if P is a 1, P′ must be 0, and vice versa. So to do this, I merely write 1s in the P′ map where there were 0s (actually blanks) in the P map. Figure 4.4-5 shows the result of this change.

Now minimize:

- Cell 3 leads to W′•X′•Y.
- Cell 14 leads to W•X.
- That leaves cells 0 and 8, which combine into X′•Y′•Z′.

**FIGURE 4.4-5:** Or–And results.

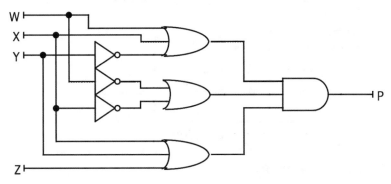

FIGURE 4.4-6: Or–And results.

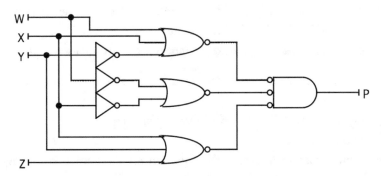

FIGURE 4.4-7: Result with double bubbles.

The result is $P' = W' \bullet X' \bullet Y + W \bullet X + X' \bullet Y' \bullet Z'$. Now that isn't quite what we want, because we need to implement the function P, not its complement.

But we have DeMorgan's Theorem (T11) that will give us the function P from the function P'. Doing this gives $P = (W + X + Y')(W' + X')(X + Y + Z)$. This arrangement requires Or gates first, connecting to an And gate on the output. The resultant circuit is shown in Fig. 4.4-6. This is the minimal Or–And form of P.

We play with the bubbles again, as shown in Fig. 4.4-7. The input Or gates have been converted to Nor gates. The output gate is the other form of a Nor gate. Figure 4.4-8 shows the final result, which is Nor–Nor gating.

Now compare the four results: And–Or (Fig. 4.4-2), Nand–Nand (Fig. 4.4-4), Or–And (Fig. 4.4-6) and Nor–Nor (Fig. 4.4-8). Which one is better? Yes! Yes, what? Yes, one of them is better. Uhuh, but which one? I don't know. How about defining "better" and we can perhaps get an answer.

Logically, all four are the same. Gate and input counts are equal, so it's hard to tell them apart.

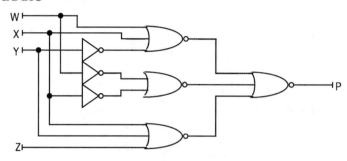

**FIGURE 4.4-8:**  Result in Nor–Nor form.

So how do you? If the boss says to use Nand–Nand, that answers the question. Sometimes, though, one simplification is clearly better than another. So sometimes And–Or and its clone Nand–Nand work out to be simpler. Sometimes not.

The steps to getting a particular arrangement of logic are easily stated:

*And–Or* gating comes from minimizing the function as given, working with the 1s of the function.

*Nand–Nand* gating comes from the And–Or result by placing bubbles on each And output and on all Or inputs.

*Or–And* gating comes from minimizing the function after complementing it, which is the equivalent of working with the 0s of the function.

*Nor–Nor* gating comes from the Or–And result by placing bubbles on each Or output and on all And inputs.

Be careful of inputs that lead directly to the output gate, though. Bubbles must be added in pairs, so such an input must be complemented.

## 4.5    DESIGN EXAMPLES

Design is more fun—at least you are creating something. Or at least that's the way I see it. I'm going to work through two examples to illustrate how a problem might be stated and then solved to produce a minimal logic circuit.

### 4.5.1    Multiplication

Let's design a multiplier circuit in Nand–Nand form that has two two-bit inputs and produces a four-bit output. Notice that the largest two-bit number is 11, which is 3. $3 \times 3 = 9$, and 9 in binary is 1001, which is four bits. Figure 4.5-1 is an example of binary multiplication.

I need variable names, so I'll follow the convention of numbering bits from the low-order end. One of the inputs will be $A_1A_0$ and the other will be $B_1B_0$. I'll label the outputs $C_3C_2C_1C_0$.

$$
\begin{array}{r}
1\ 0 \\
\mathrm{x}\quad 1\ 1 \\
\hline
1\ 0 \\
1\ 0\phantom{\ 0} \\
\hline
1\ 1\ 0
\end{array}
$$

FIGURE 4.5-1: Multiplication.

| $A_1$ | $A_0$ | $B_1$ | $B_0$ | $C_3$ | $C_2$ | $C_1$ | $C_0$ |
|---|---|---|---|---|---|---|---|
| 0 | 0 | 0 | 0 | 0 | 0 | 0 | 0 |
| 0 | 0 | 0 | 1 | 0 | 0 | 0 | 0 |
| 0 | 0 | 1 | 0 | 0 | 0 | 0 | 0 |
| 0 | 0 | 1 | 1 | 0 | 0 | 0 | 0 |
| 0 | 1 | 0 | 0 | 0 | 0 | 0 | 0 |
| 0 | 1 | 0 | 1 | 0 | 0 | 0 | 1 |
| 0 | 1 | 1 | 0 | 0 | 0 | 1 | 0 |
| 0 | 1 | 1 | 1 | 0 | 0 | 1 | 1 |
| 1 | 0 | 0 | 0 | 0 | 0 | 0 | 0 |
| 1 | 0 | 0 | 1 | 0 | 0 | 1 | 0 |
| 1 | 0 | 1 | 0 | 0 | 1 | 0 | 0 |
| 1 | 0 | 1 | 1 | 0 | 1 | 1 | 0 |
| 1 | 1 | 0 | 0 | 0 | 0 | 0 | 0 |
| 1 | 1 | 0 | 1 | 0 | 0 | 1 | 1 |
| 1 | 1 | 1 | 0 | 0 | 1 | 1 | 0 |
| 1 | 1 | 1 | 1 | 1 | 0 | 0 | 1 |

FIGURE 4.5-2: Truth table for two-bit multiplier.

This problem can be readily expressed in a truth table, which is shown in Fig. 4.5-2. I'll use K-maps to find the minimal And–Or gating (Fig. 4.5-3)

$C_3$ is rather obvious without a map—there is a single 1 in its column, so the result is
$A_1 \bullet A_0 \bullet B_1 \bullet A \bullet_0$.

$C_2$ requires two pairs, $A_1 \bullet A_0' \bullet B_1 + A_1 \bullet B_1 \bullet B_0'$.

$C_1$ is messier; it takes four pairs that won't simplify further, $A_1 \bullet B_1' \bullet B_0 + A_1 \bullet A_0' \bullet B_0 + A_0 \bullet B_1 \bullet B_0' + A_1' \bullet A_0 \bullet B_1$.

$C_0$ is a simple square, $A_0 \bullet B_0$.

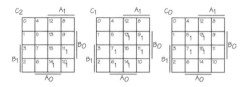

**FIGURE 4.5-3:** Two-bit multiplier maps.

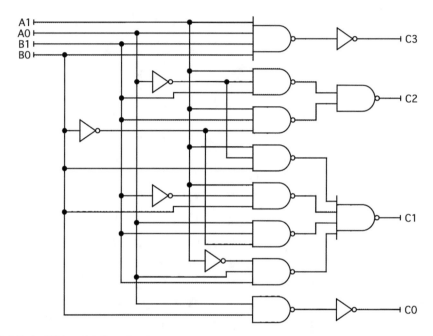

**FIGURE 4.5-4:** 25% multiplier circuit.

Figure 4.5-4 is the Nand–Nand implementation of the multiplier. The result looks complicated, but that's because of all the input lines. Notice that two of the outputs come directly from single Nand gates, so inverters are needed. While these both could have been done with And gates, the specifications said to use Nand–Nand gating. (It is not uncommon to use as an inverter a two-input Nand gate with one input connected to a 1.)

We'll see another way to do this in Chapter 5.

### 4.5.2   Flip-Flop Drivers

One of our colleagues is designing a sequential-logic circuit that has two inputs, A and B, and two output LEDs, Red and Green. If A is asserted first, then B, the output is to change to Green. The next input asserted returns the output to Red. Any other order of input events keeps the output Red.

| Current | | Input | | Next $Q_1 Q_0$ $D_1 D_0$ | | Excitation | | | |
|---|---|---|---|---|---|---|---|---|---|
| $Q_1$ | $Q_0$ | A | B | $D_1$ | $D_0$ | $J_1$ | $K_1$ | $J_0$ | $K_0$ |
| 0 | 0 | 0 | 0 | 0 | 0 | 0 | x | 0 | x |
| 0 | 0 | 0 | 1 | 0 | 0 | 0 | x | 0 | x |
| 0 | 0 | 1 | 0 | 1 | 0 | 1 | x | 0 | x |
| 0 | 0 | 1 | 1 | x | x | x | x | x | x |
| 0 | 1 | 0 | 0 | 0 | 0 | 0 | x | x | 1 |
| 0 | 1 | 0 | 1 | 0 | 1 | 0 | x | x | 0 |
| 0 | 1 | 1 | 0 | 0 | 0 | 0 | x | x | 1 |
| 0 | 1 | 1 | 1 | x | x | x | x | x | x |
| 1 | 0 | 0 | 0 | 1 | 1 | x | 0 | 1 | x |
| 1 | 0 | 0 | 1 | 0 | 0 | x | 1 | 0 | x |
| 1 | 0 | 1 | 0 | 1 | 0 | x | 0 | 0 | x |
| 1 | 0 | 1 | 1 | x | x | x | x | x | x |
| 1 | 1 | 0 | 0 | 1 | 1 | x | 0 | x | 0 |
| 1 | 1 | 0 | 1 | 0 | 1 | x | 1 | x | 0 |
| 1 | 1 | 1 | 0 | 0 | 0 | x | 1 | x | 1 |
| 1 | 1 | 1 | 1 | x | x | x | x | x | x |

FIGURE 4.5-5: Excitation table for sequential circuit.

The designer has carried out the sequential-circuit design steps, something we'll learn to do in Chapter 6. So she has the excitation table ready for us as shown in Fig. 4.5-5. When we see this, we realize this is simply a big truth table.

The designer asks us to determine whether using D flip-flops or JK flip-flops will give her a simpler circuit. She is working on a larger system that in general uses Nand gates.

Hmmm, flip-flops? So we ask her what she means by that request. She explains that she can implement her circuit using $D_1$ and $D_0$ to drive two D flip-flops, or she can use $J_1$, $K_1$, $J_0$, and $K_0$ to drive two JK flip-flops. She suspects one might have a simpler gate structure.

Aha! So we need to write the logic functions for the inputs $Q_1$, $Q_0$, A, and B in And–Or form (which leads to Nands). These are to produce the six outputs (Ds, Js, and Ks). Then we can see whether the Ds as a group are simpler than the JKs as a group.

We'll draw six maps, as shown in Fig. 4.5-6. Then we'll go through the minimization algorithm to get the following results (note the change to overbars instead of primes):

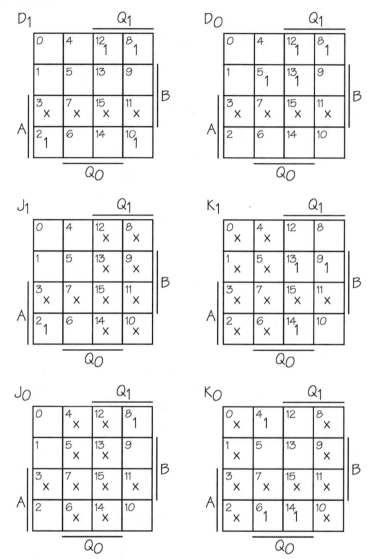

FIGURE 4.5-6: Maps for excitations.

$$D_1 = \overline{Q}_0 \bullet A + Q_1 \bullet \overline{A} \bullet \overline{B},$$
$$D_0 = Q_0 \bullet B + Q_1 \bullet \overline{A} \bullet \overline{B},$$
$$J_1 = \overline{Q}_0 \bullet A,$$
$$K_1 = B + Q_0 \bullet A,$$
$$J_0 = Q_1 \bullet \overline{A} \bullet \overline{B},$$
$$K_0 = A + \overline{Q}_1 \bullet \overline{B}.$$

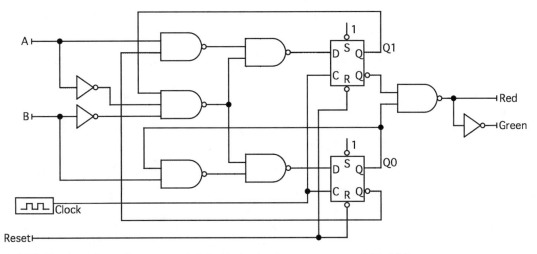

**FIGURE 4.5-7**: Complete sequential circuit (excitations are gates on left side).

Now let's compare the results. The two D outputs require 5 gates and 11 inputs (one Nand gate is common to both $D_1$ and $D_0$). The four JK outputs require 6 gates and 13 inputs.

We'll recommend the D solution. Figure 4.5-7 shows the entire sequential circuit that our colleague has designed by following our recommendation. She had already developed the logic for the output:

$$Green = \overline{Q_1} \bullet Q_0,$$

$$Red = \overline{Green}.$$

My choice of Ds conforms to the fact that most designs today use only D flip-flops. This is because of the manner in which large logic circuits are instantiated on programmable logic devices on a chip.

## 4.6 SUMMARY

Did you get to this point feeling like a lot has happened? It has! We've covered all the basics of combinational design, at least for modest-sized logic circuits.

We started with gates—And, Or, Not, Exclusive Or, and Nand and Nor. The last two are most common. Then we tried some analysis before heading into design and minimization.

The Karnaugh map minimization technique works well for "people" if the circuit isn't too large. Visualizing pairs leads to minimal gate and input counts. Don't-cares often reduce things even more.

In the next chapter, we'll see that there are larger building blocks that can improve the design process even more.

CHAPTER 5

# Building Blocks: Bigger Stuff

Lots of detail! Does a "real" designer use all these little devices like individual gates and inverters? If so, it must take a century to design a computer. You know the answer—things come in bigger packages. So big that modern processors have millions of logic elements in one package.

No, we aren't going that far yet. It would help to know how to walk briskly before we try to run. So what we are looking at in this chapter are "bigger" devices. These integrate into one package a number of gates and other logic elements. Still not very many, perhaps as many as 30, but an improvement over just four Nand gates in a 74LS00.

Historically, this integration proceeded at about the same pace as the integrators. As fast as the silicon designers could make things smaller, the logic designers wanted more things on a chip. The integration of four Nand gates into a single chip was a major advance in its time. The integration of 30 gates into an arithmetic and logic unit was a major advance for its time too. And for that matter, a computer on a chip was . . . .

In this chapter, we'll work with just a few of the "bigger" devices, such things as decoders, multiplexers, devices for arithmetic, read-only memories, and programmable logic arrays. We are also going to encounter the first of a number of "problems" that arise when designing logic: hazards.

## 5.1   DECODER

Decoders and multiplexers have names that describe the logical functions that they are intended to perform. But they will do other things as well, and it's in the "other things" mode that we'll use them. What they do is give us a way of developing a logic function with just one part. So far we've been implementing functions with groups of gates.

The decoder shown in Fig. 5.1-1 has four outputs. One of them is made active by selecting it with the two S inputs. Since the two inputs can together have any one of four binary values (S1 S0 = 00, 01, 10, or 11), this gives us a way to select, using a *code*, one of the four inputs. Hence, this device is called a *2-to-4 decoder*.

The enable line, labeled EN, "enables" the entire device. Notice that the bubble tells us that EN must be a 0 to enable. So if EN = 1, no output can be selected, no matter what the S lines are saying. The outputs are all 0.

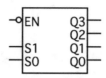

FIGURE 5.1-1:  2-to-4 decoder.

| EN | I1 | I0 | Q3 | Q2 | Q1 | Q0 |
|----|----|----|----|----|----|----|
| 1  | x  | x  | 0  | 0  | 0  | 0  |
| 0  | 0  | 0  | 0  | 0  | 0  | 1  |
| 0  | 0  | 1  | 0  | 0  | 1  | 0  |
| 0  | 1  | 0  | 0  | 1  | 0  | 0  |
| 0  | 1  | 1  | 1  | 0  | 0  | 0  |

FIGURE 5.1-2:  Reduced truth table for 2-to-4 decoder.

The truth table of Fig. 5.1-2 shows how this device works. But this is an incomplete truth table, at least compared to what we have been seeing. I've reduced it by combining all four rows that have $EN = 1$. After all, the outputs are all going to be 0, no matter what is on the S lines.

You'll see this reduced truth table quite often, especially in manufacturer's literature. It saves lots of space, especially for more complex devices. I've used the "x" for don't-care outputs. Here, it means a don't-care input.

This is a good point to stray for a moment. Have you noticed that enable inputs tend to be active low? That they tend to have a bubble? Why? I don't know all the reasoning, but I have a guess. In standard transistor–transistor logic (TTL), which is what the 74LS series is made out of, inputs that are not connected to anything tend to float to the high logic level. If it takes a low logic level to enable the chip, then the chip cannot be enabled by accident if we forget to provide an enable or if the enable becomes disabled. Another reason may be that the transistor circuit is slightly simpler and possibly faster. But whatever the reason, these inputs tend to be upside down.

### 5.1.1   Minterm Generator

The decoder clearly can receive a binary code and break that code out into one of several outputs. But for us its major use will be as a *minterm generator*. Why do we need minterms? Remember that these come from the truth table. We've been using Karnaugh maps to reduce the minterms to "smaller" combinations, then designing And–Or circuits to provide them.

The decoder can generate minterms. For example, the output $Q2 = EN' {\bullet} I1 {\bullet} I0'$. So if we enable the chip by making $EN = 0$, then $Q2 = I1 {\bullet} I0' = \Sigma(2)$. And there's a minterm!

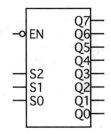

FIGURE 5.1-3:  3-to-8 decoder.

R                              F

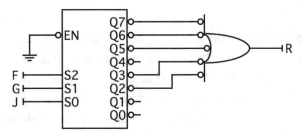

FIGURE 5.1-4:  Map of R.

FIGURE 5.1-5:  Decoder implementation.

Let's apply this knowledge using a 3-to-8 decoder shown in Fig. 5.1-3. Implement $R(F,G,J) = \Sigma(2,3,5,6,7)$. If we were doing this in Chapter 4 manner, we'd draw a map (Fig. 5.1-4) and minimize the function to $R = G + F \bullet J$. Then we'd design a gate circuit consisting of one And gate and one Or gate.

I can do this with the 3-to-8 decoder and one Or gate as well. Figure 5.1-5 shows the result. Did I save any logic? Not really, for it still takes two devices, the decoder and the Or gate. But I have simplified the design. And sometimes, especially in "disorderly" designs where the logic doesn't simplify much, the decoder will be simpler too.

Notice the order of the inputs. In the function given, the order is specified by $R(F,G,J)$. In other words, F is the high-order bit when we calculate the numerical values of the minterms. Hence, F is assigned to the S2 selection input because S2 is the high-order bit of that input.

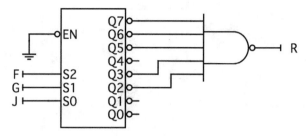

FIGURE 5.1-6:  Same implementation.

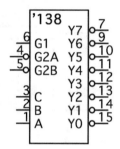

FIGURE 5.1-7:  74LS138 decoder.

In general, inputs and outputs that have some order to them are numbered from "0," which is the low-order bit of the code. Sometimes you'll find that inputs have letters instead. As a general rule, the input labeled "A" is the low-order bit of the code.

Commonly available decoders have inverted outputs. In other words, the output is active when it is low. So a 0 output is the active output. But as Fig. 5.1-6 shows, this doesn't change things much for us. The implementation of R(F,G,J) now requires complemented inputs to the Or gate. We recognize this as a Nand gate.

Larger decoders can be made from smaller ones. Let's take our 3-to-8 decoder and make a 4-to-16 decoder. But this time I'll do it with a commercial part, the 74LS138 3-to-8 decoder. Figure 5.1-7 shows the '138. Notice that its outputs are inverted. But more important are the three enables, labeled "G" here (to stand for Gate inputs). Two are active low and one is active high. This makes the commercial device more flexible. After all, if you are going to market something, it's nice to be able to market it for many different purposes.

By combining the inputs and the enables properly, I can make a 4-to-16 decoder easily. If my inputs are to be D, C, B, and A (high to low), I will use C, B, and A to drive the selection inputs of two '138 decoders. Figure 5.1-8 shows the circuit.

The input D is assigned to choose between the chips via the enable inputs. I connect D to the active-high enable on the upper chip to provide the eight high-order outputs. I connect D to an active-low enable on the lower chip to provide the eight low-order outputs.

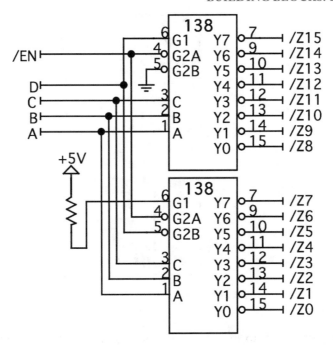

FIGURE 5.1-8: 4-to-16 decoder.

In keeping with active-low enables, my enable is active low. So I drive an active-low input on each device. The remaining enable inputs must be made active by grounding active-low enables and pulling up active-high inputs.

Notice my labels. Since the output Z15 is active low, I write /Z15. Similarly, my enable is active low and is labeled /EN. But when I draw a single block (Fig. 5.1-9) to show my new device, I use Z15 and EN *inside*, because the complementing happens through the bubble.

## 5.1.2  Two Applications of Decoders

Remember the two-bit multiplier of Section 4.5-1? The truth table is shown in Fig. 4.5-2. The functions are

$$C_3(A_1, A_0, B_1, B_0) = \sum(15),$$

$$C_2(A_1, A_0, B_1, B_0) = \sum(10, 11, 14),$$

$$C_1(A_1, A_0, B_1, B_0) = \sum(6, 7, 9, 11, 13, 14),$$

$$C_0(A_1, A_0, B_1, B_0) = \sum(5, 7, 13, 15).$$

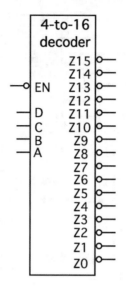

FIGURE 5.1-9:  1-9 Decoder representation.

The *random-logic* circuit to implement this multiplier is shown in Fig. 4.5-4. Here is a parts list using standard 74LS series parts:

$$
\begin{array}{ll}
6 \text{ inverters} & = 174LS04 \\
2 \text{ 2-in Nand gates} & = 1\ 74LS00\ (2\ \text{extra}) \\
6 \text{ 3-in Nand gates} & = 2\ 74LS10 \\
2 \text{ 4-in Nand gates} & = 1\ 74LS20
\end{array}
$$

That's a total of five parts.

Now try this with a decoder. I need minterms from 0 to 15, so I need a 4-to-16 decoder. Let's just use the one we built up in the previous section.

C3 requires an inverter to invert the active-low output. C2, C1, and C0 need Or gates with inverted inputs, which are Nand gates. The circuit is shown in Fig. 5.1-10. Is this an improvement? Let's make a parts count:

$$
\begin{array}{ll}
2 \text{ 3-to-8 decoders} & = 2\ 74LS138 \\
1 \text{ inverter} & = 1\ 74LS04\ (5\ \text{extra}) \\
1 \text{ 3-in Nand gates} & = \text{use extra 4-in Nand} \\
1 \text{ 4-in Nand gates} & = 1\ 74LS20 \\
1 \text{ 8-in Nand gate} & = 1\ 74LS30
\end{array}
$$

That's a total of five parts. Hmmm, didn't win on that one, did I! But the circuit itself is somewhat simpler because the wiring is less complex. So I have gained something. Also, this

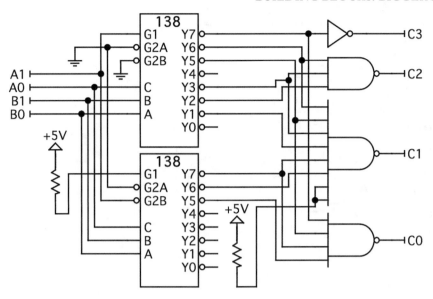

**FIGURE 5.1-10:** 10-25 multiplier again.

decoder implementation is going to be a little faster in operation, i.e., signals will pass from input to output a little more rapidly. That is sometimes important. (This can be done using the 4-to-16 decoder 74LS154, which will reduce the parts count.)

Let's try another one that has messy logic. In the sequential circuit of Fig. 4.5-5 we used random logic to implement the functions that drive the two flip-flop inputs. Perhaps a decoder can help?

The functions we need are

$$D_1(Q_1, Q_0, A, B) = \sum (2, 8, 10, 12),$$

$$D_0(Q_1, Q_0, A, B) = \sum (5, 8, 12, 13).$$

Figure 4.5-7 shows the random logic. Don't-cares don't help here. The parts count is

2 D flip-flops   = 1 74LS175 (2 extra)

5 2-in Nand gates = 1 74LS00 (1 from 74LS10)

1 3-in Nand gate   = 1 74LS10 (2 extra)

3 inverters   = 1 74LS04 (3 extra)

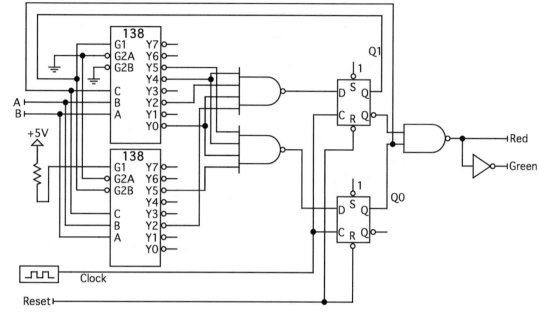

**FIGURE 5.1-11:** Sequential circuit again.

That's four parts. Can the decoder design do better? Figure 5.1-11 shows the decoder design. The logic doesn't look quite as complicated, but how about the parts count?

$$
\begin{array}{ll}
\text{2 D flip-flops} & = \text{1 74LS175 (2 extra)} \\
\text{2 3-to-8 decoders} & = \text{2 74LS138} \\
\text{12-in Nand gate} & = \text{174LS00 (3 extra)} \\
\text{24-in Nand gates} & = \text{1 74LS20} \\
\text{2 inverters} & = \text{use extra 2-in Nands}
\end{array}
$$

That's five parts. Nope, it's worse!

### 5.1.3   Other Decoders

There are other decoders around that really do decode! There is a group of decoders that convert binary-coded-decimal (BCD) signals into seven-segment displays. BCD is common because we often want to display numbers in friendly form. We can convert binary results into BCD codes, so we just need a decoder to convert the code into the proper segments.

One such decoder is the 7447, which receives a BCD code and provides the seven signals for the seven segments. Figure 5.1-12 shows the decoder and a seven-segment display.

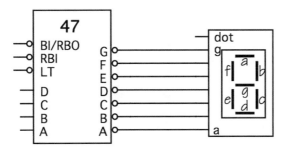

FIGURE 5.1-12: Seven-segment decoder and display.

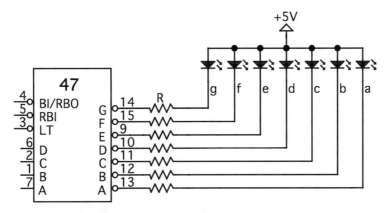

FIGURE 5.1-13: Seven-segment electrical connevtions.

The '47 has several features. First, it has enough oomph to drive an LED to about 25 milliamperes, bright enough to be useful. Second, it can be cascaded into a group of digits that will automatically suppress leading zeros. (BI is the blanking input, RBO is the ripple-blanking output, and RBI is the ripple-blanking input. Combining these makes the zero suppression work. Their use is shown in Fig. 8.3-5.) Third, it has provisions (LT) for lighting all the LEDs to make sure they all work.

Figure 5.1-13 shows one implementation. The value of the resistor R is determined by the maximum current that is to pass through the LED.

## 5.2    MULTIPLEXER

You can look at decoders and multiplexers sort of like you look at a funnel. The decoder is like looking into the small end of the funnel (the code inputs) and seeing out the large end (the selectable output lines). The multiplexer is the opposite. The big end of the funnel is the input lines, one of which is selected to appear as the output.

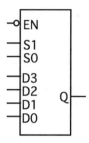

FIGURE 5.2-1: 4-to-1 multiplexer.

| EN | S1 | S0 | Q |
|----|----|----|----|
| 1 | x | x | 0 |
| 0 | 0 | 0 | D0 |
| 0 | 0 | 1 | D1 |
| 0 | 1 | 0 | D2 |
| 0 | 1 | 1 | D3 |

FIGURE 5.2-2: Multiplexer truth table.

Basically, the multiplexer selects an input from one of its lines to become the output. Figure 5.2-1 is a 4-to-1 multiplexer. The two selection lines (S) choose one of the four data lines (D) and connect it to the output. (I will get tired of typing and start using the shorter term *mux*, which is quite common.)

The truth table for the 4-to-1 mux is shown in very reduced form in Fig. 5.2-2. This is a huge reduction in the "standard" form of a truth table. After all, the mux has seven inputs, so the full table would have $2^7 = 128$ rows. This one has only five.

Notice the active-low enable again. Unless it is 0, there's no output. (Well, the output is 0.) When it's 0, the output is selected by the S lines.

Larger muxes are available. Figure 5.2-3 shows an 8-to-1 mux. It has two outputs, one the complement of the other. Also common are chips with several smaller muxes on one chip.

What's a mux for? It's basic purpose is to select among several signals and pass just one through. To multiplex signals means to place several on a single transmission line. When you talk on the telephone, your voice is digitized and transmitted on the same circuit as many other voices. The various voices are multiplexed onto the circuit at one end and demultiplexed (demuxed) at the other. Obviously, there is some mechanism for identifying which ones go with which calls.

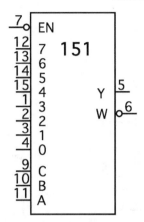

FIGURE 5.2-3:  8-to-1 Multiplexer.

## 5.2.1   One-Chip Designs

Muxes can provide "one-chip" design for simple logic too, just as decoders can. Since the mux can produce only one output (and perhaps its complement), it takes one mux for each signal in the design. But this can sometimes save chips in the design of random logic.

Suppose we want $G(A,T,X) = \Sigma(0,2,5,6,7)$. There are three input bits (A, T, and X) so I can use these to select any one of eight inputs. I'll choose each input to be 0 or 1 depending on the function. For A = 0, T = 0, X = 0, the input should be 1 so that the output for the $\Sigma(0)$ term will be 1 as specified by the function G.

Figure 5.2-4 shows implementation with an 8-to-1 mux. I get the necessary inputs using ground, which is 0, and +5 volts with a pull-up resistor, which is 1. The enable line is grounded to permanently enable the mux.

But suppose I need a function of four inputs such as $K(A,B,C,D) = \Sigma(3,4,6,7,9,10,11,14)$. This has one more input than I have select lines on an 8-to-1 mux. This can be done by partitioning the truth table for the function into pairs of inputs.

Figure 5.2-5 shows the truth table for my function K. I choose to use A, B, and C as the three selection inputs to my 8-to-1 mux. That leaves D as the odd input. I write the table so that D is the rightmost column (which it was already). Then I divide the table into pairs of D inputs, 0 and 1. Now look at K in each of the partitions. K has one of four "values" in each partition: 00, 01, 10, or 11. This output relates to D in one of four ways:

00   the output is 0 no matter what D is;
01   the output is the same as D;
10   the output is the complement of D; and
11   the output is 1 no matter what D i1s.

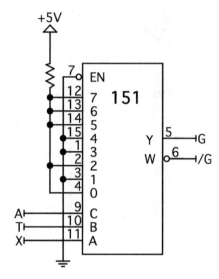

FIGURE 5.2-4:  Mux implementation of G(A,T,X).

| A | B | C | D | K | Input |
|---|---|---|---|---|-------|
| 0 | 0 | 0 | 0 | 0 | 0 = 0 |
| 0 | 0 | 0 | 1 | 0 |       |
| 0 | 0 | 1 | 0 | 0 | 1 = D |
| 0 | 0 | 1 | 1 | 1 |       |
| 0 | 1 | 0 | 0 | 1 | 2 = D' |
| 0 | 1 | 0 | 1 | 0 |       |
| 0 | 1 | 1 | 0 | 1 | 3 = 1 |
| 0 | 1 | 1 | 1 | 1 |       |
| 1 | 0 | 0 | 0 | 0 | 4 = D |
| 1 | 0 | 0 | 1 | 1 |       |
| 1 | 0 | 1 | 0 | 1 | 5 = 1 |
| 1 | 0 | 1 | 1 | 1 |       |
| 1 | 1 | 0 | 0 | 0 | 6 = 0 |
| 1 | 1 | 0 | 1 | 0 |       |
| 1 | 1 | 1 | 0 | 1 | 7 = D' |
| 1 | 1 | 1 | 1 | 0 |       |

FIGURE 5.2-5:  Truth table for K(A,B,C,D) partitioned on D.

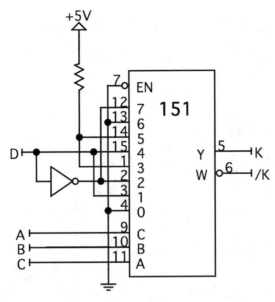

FIGURE 5.2-6:  Mux implementation of K(A,B,C,D).

I've marked the truth table to reflect these choices. Now, what does the table say? Consider the first pair of rows. A = 0, B = 0, and C = 0 are the inputs to the S lines of the mux. They will select the input line to connect to the output. If I assign a 0 to that input line, the output will be 0 for this combination of A, B, and C, for both D = 0 and D = 1.

In the next row, A =0, B = 0, and C = 1. This selects the "1" input line, which I'll connect to the signal D. Then when this combination of A, B, and C comes up, the output is the same as the input signal D.

Check the next two pairs of the table and you'll see the remaining two input combinations. The circuit is shown in Fig. 5.2-6. It requires one inverter (to get D'), so this implementation takes two parts, the mux and the inverter. (This function can be implemented without an inverter by choosing a different input as the "rightmost column," something I'll leave for you to try. Here's a hint: write the inputs in the order ABDC and then resort the truth table into proper binary order.)

## 5.2.2  Two Applications of Multiplexers

The two-bit multiplier that I have now done twice can be done a third time using a multiplexer. I'll skip some of the preliminaries and go directly to the truth table in Fig. 5.2-7. I've used the inputs in the usual order, which means that I am going to make B0 the "rightmost column" and divide the table into pairs.

| A1 | A0 | B1 | B0 | C3 In | | C2 In | | C1 In | | C0 In | |
|----|----|----|----|----|----|----|----|----|----|----|----|
| 0 | 0 | 0 | 0 | 0 | 0 | 0 | 0 | 0 | 0 | 0 | 0 |
| 0 | 0 | 0 | 1 | 0 | | 0 | | 0 | | 0 | |
| 0 | 0 | 1 | 0 | 0 | 0 | 0 | 0 | 0 | 0 | 0 | 0 |
| 0 | 0 | 1 | 1 | 0 | | 0 | | 0 | | 0 | |
| 0 | 1 | 0 | 0 | 0 | 0 | 0 | 0 | 0 | 0 | 0 | B0 |
| 0 | 1 | 0 | 1 | 0 | | 0 | | 0 | | 1 | |
| 0 | 1 | 1 | 0 | 0 | 0 | 0 | 0 | 1 | 1 | 0 | B0 |
| 0 | 1 | 1 | 1 | 0 | | 0 | | 1 | | 1 | |
| 1 | 0 | 0 | 0 | 0 | 0 | 0 | 0 | 0 | B0 | 0 | 0 |
| 1 | 0 | 0 | 1 | 0 | | 0 | | 1 | | 0 | |
| 1 | 0 | 1 | 0 | 0 | 0 | 1 | 1 | 0 | B0 | 0 | 0 |
| 1 | 0 | 1 | 1 | 0 | | 1 | | 1 | | 0 | |
| 1 | 1 | 0 | 0 | 0 | 0 | 0 | 0 | 0 | B0 | 0 | B0 |
| 1 | 1 | 0 | 1 | 0 | | 0 | | 1 | | 1 | |
| 1 | 1 | 1 | 0 | 0 | B0 | 1 | B0' | 1 | B0' | 0 | B0 |
| 1 | 1 | 1 | 1 | 1 | | 0 | | 0 | | 1 | |

FIGURE 5.2-7: Truth table for two-bit multiplier partitioned.

There are four outputs to implement, so I need four muxes. The result is the circuit in Fig. 5.2-8. It needs an inverter, which means I need five parts (four muxes and one inverter). But that's the same number that I needed in the original circuit and in the decoder implementation. So did this improve anything? Probably not, if for no other reason than this is a somewhat more confusing circuit.

Can this multiplier be done without an inverter? By choosing a different "rightmost column," there's a chance, but don't be too optimistic!

The sequential circuit that I have done twice is another good test of this mux idea. The truth table is shown in Fig. 5.2-9. I have chosen the B input as the "rightmost column." The circuit that results is shown in Fig. 5.2-10. This takes three muxes, two for the Ds and one for the output. After all, if muxes are this good, let's us 'em! But the circuit still takes five parts (3 muxes, 1 pair of flip-flops, and an inverter); random logic can be done in four.

By being clever, this can be done in just four parts—the inverter can be avoided. This requires a proper choice of the "rightmost column" and another less obvious step. Sounds like a good problem!

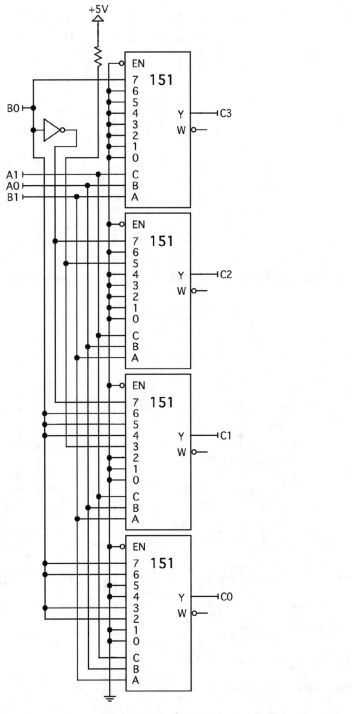

FIGURE 5.2-8: Two-bit multiplier using multiplexers.

| Q1 | Q0 | A | B | D1 In | | D0 In | | Red In | |
|----|----|---|---|-------|----|-------|----|--------|----|
| 0 | 0 | 0 | 0 | 0 | 0 | 0 | 0 | 0 | 0 |
| 0 | 0 | 0 | 1 | 0 | | 0 | | 0 | |
| 0 | 0 | 1 | 0 | 1 | 1 | 0 | 0 | 0 | 0 |
| 0 | 0 | 1 | 1 | 1 | | 0 | | 0 | |
| 0 | 1 | 0 | 0 | 0 | 0 | 0 | B | 1 | 1 |
| 0 | 1 | 0 | 1 | 0 | | 1 | | 1 | |
| 0 | 1 | 1 | 0 | 0 | 0 | 0 | 0 | 1 | 1 |
| 0 | 1 | 1 | 1 | 0 | | 0 | | 1 | |
| 1 | 0 | 0 | 0 | 1 | B' | 1 | B' | 0 | 0 |
| 1 | 0 | 0 | 1 | 0 | | 0 | | 0 | |
| 1 | 0 | 1 | 0 | 1 | 1 | 0 | 0 | 0 | 0 |
| 1 | 0 | 1 | 1 | 1 | | 0 | | 0 | |
| 1 | 1 | 0 | 0 | 1 | B' | 1 | 1 | 0 | 0 |
| 1 | 1 | 0 | 1 | 0 | | 1 | | 0 | |
| 1 | 1 | 1 | 0 | 0 | 0 | 0 | 0 | 0 | 0 |
| 1 | 1 | 1 | 1 | 0 | | 0 | | 0 | |

FIGURE 5.2-9: Truth table for sequential circuit functions partitioned.

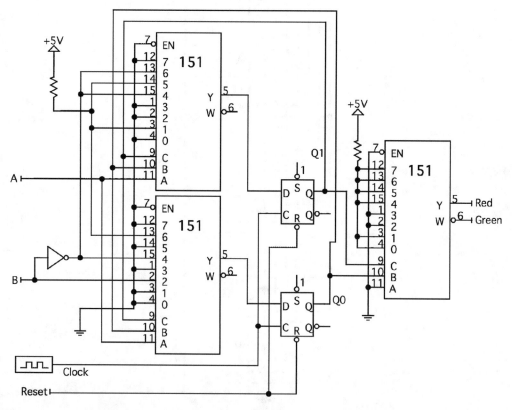

FIGURE 5.2-10: Sequential circuit using multiplexers.

## 5.3   ARITHMETIC

There are lots of "bigger chips" available. The decoder and the multiplexer are just two examples. I'll finish this look at some by considering arithmetic. This is something that we often want to do, so it isn't a surprise that folks have designed devices for doing it.

Start with the simplest arithmetic, the addition of two bits in binary. The truth table for this addition in Fig. 5.3-1 shows the output sum and the output carry. This is a good example of a circuit that uses an Exclusive-or gate very well. The circuit is shown in Fig. 5.3-2.

This adder is called a *half adder*, though, because it can add only two bits. If we are going to add binary numbers, we need to handle carry bits as well. This means three inputs: the two bits to sum and the carry from the previous sum. Its truth table is shown in Fig. 5.3-3. This leads to the circuit shown in Fig. 5.3-4, which has six gates and 13 inputs.

| A | B | S | Co |
|---|---|---|---|
| 0 | 0 | 0 | 0 |
| 0 | 1 | 1 | 0 |
| 1 | 0 | 1 | 0 |
| 1 | 1 | 0 | 1 |

FIGURE 5.3-1: Half adder.

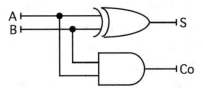

FIGURE 5.3-2: Half adder.

| A | B | Ci | S | Co |
|---|---|---|---|---|
| 0 | 0 | 0 | 0 | 0 |
| 0 | 0 | 1 | 1 | 0 |
| 0 | 1 | 0 | 1 | 0 |
| 0 | 1 | 1 | 0 | 1 |
| 1 | 0 | 0 | 1 | 0 |
| 1 | 0 | 1 | 0 | 1 |
| 1 | 1 | 0 | 0 | 1 |
| 1 | 1 | 1 | 1 | 1 |

FIGURE 5.3-3: Full adder.

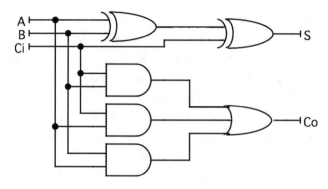

FIGURE 5.3-4:  Full adder: 6-gate arrangement.

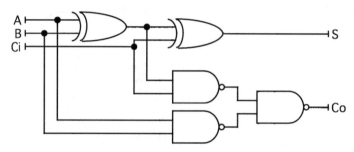

FIGURE 5.3-5:  Full adder: 5-gate arrangement.

A circuit with five gates and 10 inputs is shown in Fig. 5.3-5. Is this an improvement? Well, sure, it has fewer parts, right? But how about the time for a signal to make it through the circuit? What is the longest path in each? The second circuit also has a problem called a *glitch* that we will deal with later in this chapter.

The carry operation in addition takes time. Consider the addition of two four-bit numbers. You add the rightmost two bits and produce a carry. *Then* you can add the next two bits and the carry you just produced. *Then* you can do the next bits to the left. *Finally* you can add the leftmost bits. So a four-bit addition isn't finished until the carry from the right end has had time to propagate all the way to the left. This is called the *carry ripple* and it takes time.

The ripple problem becomes very large when you are building a microprocessor that adds two 64-bit numbers! If we had to wait for the carry, microprocessors would be pretty slow. But the carry can be predicted by a combinational circuit. In other words, the carry at the left end of a four-bit addition can be calculated without waiting. This is called *carry look-ahead*. There are chips that do this.

Addition is not the only computation we might want, so there are larger devices that do more. One group of these is called *arithmetic-logic unit*, which gets abbreviated ALU. These can typically do more than just addition.

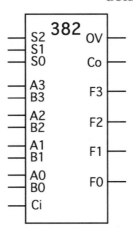

FIGURE 5.3-6: Arithmetic-logic unit.

| S2 | S1 | S0 | Output function |
|----|----|----|-----------------|
| 0  | 0  | 0  | All 0           |
| 0  | 0  | 1  | B–A–1+Ci        |
| 0  | 1  | 0  | A–B–1+Ci        |
| 0  | 1  | 1  | A+B+Ci          |
| 1  | 0  | 0  | A xor B         |
| 1  | 0  | 1  | A or B          |
| 1  | 1  | 0  | A and B         |
| 1  | 1  | 1  | All 1           |

FIGURE 5.3-7: '382 functions.

One example is the four bit ALU 74LS382 (Fig. 5.3-6). The select (S) inputs select the function to be performed on the two four-bit input numbers. The table in Fig. 5.3-7 shows the eight functions.

Look at the table and find simple four-bit addition. Yup, the 011 line, $A + B + C_i$ to include an incoming carry. Notice that $C_o$ is the outgoing carry after the four-bit addition. The device employs carry look-ahead to speed things up.

But what are all those other functions. Well, I think the logic operations are obvious: all 0s, Exclusive-or, Or, And, and all 1s. But what are the lines 001 and 010? These do twos-complement subtraction and properly include the carry, which really is the borrow. The device's OV output line provides an indication of twos-complement overflow. We aren't going any further with this!

## 5.4    PROGRAMMABLE BLOCKS

Sometimes it's nice to have a general logic device that we can program for some specific function. Much random logic today is replaced by these programmable devices because they can fit a large amount of functionality into one chip. Also, the logic can be changed by programming and these devices can do more at less cost.

These devices include read-only memory (ROM) in various forms and programmable logic devices (PLD). We are going to look at just a couple of examples of these so that you know a little about their existence.

### 5.4.1   Read-Only Memory

The basic idea of memory is, of course, a place to store bits or groups of bits. Memory consists of address lines in and data lines in or out. The address lines select the place to store bits or to find and retrieve them. The data lines are the paths these bits move along.

Read-only memory (ROM) can't store new bits; it can only give them back. The bits have to be stored originally, of course, and how this is done distinguishes one form of ROM from another:

ROM is the basic form of read-only memory. The bits are stored permanently in the memory at the time the device is made.

PROM is Programmable ROM. The memory is made with all the bits set to one value (usually 1). The user then uses a *programmer* to set up the bits in the pattern desired. Once this is done, the memory is permanent.

EPROM is Erasable PROM. The bits can be restored to one common value using ultraviolet light. Then the user can use a programmer to set up the desired values. This device can be reused numerous times.

EEPROM is Electrically Erasable PROM. The bits can be changed under special conditions in the circuit, but they require no power to maintain their values.

No matter which form of ROM we are talking about, they all have the same function—to store bits. These bits can be program instructions to provide a permanent program for a microcomputer. The program that runs first when a PC is turned on comes from ROM. So in that sense the ROM is nothing more than a permanent computer memory.

But the ROM can also be thought of as a truth table in electrical form. It's in that form that we will use it. In other words, we can use a ROM to provide a logic function by simply putting into it the bits of the truth table we want.

How do we describe a ROM? Well, it has address lines in and data lines out. So we describe it by stating the number of addresses it has and then the number of data lines. For example, a ROM with eight address lines and eight data lines would be called a 256-by-8 ROM, or 256×8. The eight data lines can address $2^8 = 256$ different groups of eight bits. Hence, this ROM can mechanize a truth table with eight inputs (and hence 256 rows) and eight output columns.

Here's a simple example, certainly not with a 256-row truth table! Let's implement our two-bit multiplier in a ROM. There are four inputs, A1, A0, B1, and B0, so I need four address lines. There are four outputs, C3, C2, C1, and C0, so I need four output lines. The ROM is therefore 16×4 because $2^4 = 16$. That's a pretty tiny ROM—one this small doesn't exist commercially.

We often tabulate the truth table for a ROM in a fatter array than we have used before. Figure. 5.4-1 shows the multiplier truth table with two inputs down the left side and two across the top. The four outputs are given as groups of four bits in the tabulation. Figure 5.4-2 shows the circuit drawing of the ROM itself. (Common ROM devices also have an enable input.)

## 5.4.2    Programmable Logic Device

Generic programmable logic is called a PLD, programmable logic device. Very simply, this is a logic device whose function is not built into it. It's up to the user to program it to have the desired function.

PLDs come in variations similar to ROM, programmable, erasable, and so on. They go by various names that are often trademarks of their manufacturers.

| A1 A0 | B1 B0 | | | |
|-------|-------|------|------|------|
|       | 00 | 01 | 10 | 11 |
| 0  0 | 0000 | 0000 | 0000 | 0000 |
| 0  1 | 0000 | 0001 | 0010 | 0011 |
| 1  0 | 0000 | 0010 | 0100 | 0110 |
| 1  1 | 0000 | 0011 | 0110 | 1001 |

FIGURE 5.4-1: ROM for multiplier.

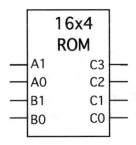

FIGURE 5.4-2: ROM.

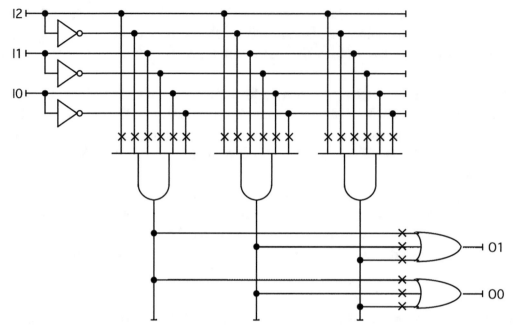

FIGURE 5.4-3: 3×2 programmable logic array (PLA).

The simplest of the PLDs is the programmable logic array, or PLA. This device is nothing more than a collection of And–Or logic. In this collection, *every* input and its complement goes to each And gate. The output of *every* And gate goes to each Or gate. To program the device, you break some of these connections.

The PLA is described by giving the number of inputs and the number of outputs. I've shown a 3-by-2 (3×2) PLA in Fig. 5.4-3. You can see that every And gate has six inputs, one for each input and its complement. Every Or gate has three inputs, one from each And gate. The "X" symbol means that there is a "fuse" at that point which can be broken by programming.

Suppose we want to program this simple PLA to implement O1 = I2•I1•'I0, O0 = I2•I0' +I1. The result is the links shown in Fig. 5.4-4.

What's the difference between this and ROM? Consider my simple 3×2 example. First, ROM remembers; this PLA has no memory in the computer sense. Second, if a ROM has three inputs (three address lines) it must have eight different places to store bits. My PLD example has only three—the three vertical And gates.

PLDs can be much more complex. One common family has the same arrangement of And gates but the connections to the Or gates are not programmable. Another family includes the And and Or gating and then adds flip-flops. Others include internal memory in the form of

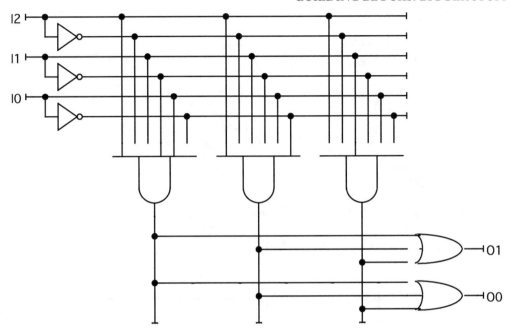

FIGURE 5.4-4: Programmed 3×2 PLA.

static RAM (read–write memory that remembers with the power off). This memory determines the function of the device.

## 5.5 HAZARDS AND GLITCHES

A *glitch* is an unwanted pulse that appears in an output that is supposed to be steady. Huh? Well, a glitch is something you don't want in a signal but it's there anyhow. Hmmm, no, that doesn't say it either.

Suppose the output of a certain logic circuit is supposed to stay 1. Suppose further that a certain input change is not supposed to change the output. If the output changes briefly during that input change, that's a glitch.

Are glitches bad? No, unless they have some bad effect on the circuit downstream. For example, a glitch could trigger a flip-flop and make it change when it isn't supposed to. That's a glitch that you don't want.

### 5.5.1 Example of Glitch

I can demonstrate a glitch in the five-gate version of the full-adder circuit that I showed in Fig. 5.3-5, redrawn here as Fig. 5.5-1. The glitch occurs when $B = 1$ and $Ci = 1$. No matter what A is, the carry output Co should stay 1, because $1 + 1 +$ anything yields a carry.

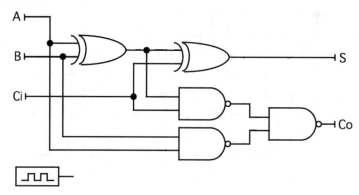

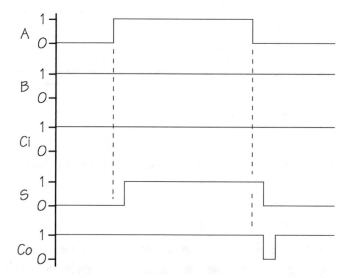

FIGURE 5.5-1:  5-gate full adder for glitch testing.

FIGURE 5.5-2:  Adder timing with glitch.

If $A = 1$ and then is changed to 0, the carry output goes to 0 briefly. The timing is shown in Fig. 5.5-2. You can see the little pulse just after A changes from 1 to 0. This is the glitch.

Is this a bad glitch? Only if the circuit downstream is affected by it. Sometimes it's OK to leave glitches in the circuit because they don't do anything wrong. but generally they shouldn't be there, especially in sequential circuits where time is a factor.

Where did this glitch come from? This full-adder implementation is three-level logic. But not all signals go through all three levels. The problem is that both A and B arrive at Co through both a two-level path (the bottom Nand gates) and a three-level path (the first Exclusive-or gate). These differently timed arrivals affect the carry output individually, making the glitch.

How do we prevent glitches? Or better, how do we design circuits that will not have glitches? One answer is somewhat obvious—all signal paths must be the same length. Another answer is to keep things simple—use two-level logic only. But neither of these is easy to implement as such.

There's another approach. To get there, I need to define a *hazard*. A hazard is a condition in a logic circuit that *could* produce a glitch. If our circuit is free of hazards, it can't produce any glitches.

## 5.5.2   Hazards

There are two kinds of hazards in logic circuits, not counting the Dukes of: static and dynamic

A *static hazard* is a situation where an output is supposed to stay at a particular value during an input change but there's a chance it won't (possibly yielding a glitch). So a *static-1* hazard is one where the output is supposed to remain 1 during an input change.

Notice that a static hazard doesn't have to actually produce a glitch during an input change. There's only the possibility of a glitch.

A *dynamic hazard* is more complicated. An input change is supposed to change the output. But the output has the possibility of not changing smoothly. For example, suppose an output is to change from 1 to 0 due to a particular input change. Instead, it changes from 1 to 0 to 1 to 0. That's clearly wrong and that's what a dynamic hazard can do.

Avoiding dynamic hazards is very simple—don't design logic beyond two levels. That's all there is to it, and that's what we've been doing anyhow. So I can ignore dynamic hazards from now on, and so can you, because we'll stick with two-level logic.

With those out of the way, how do we avoid static hazards? Half the answer is simple. If we are using And–Or or Nand–Nand logic, we are designing with the 1s of the output in mind. Hence, the circuit cannot have any static-0 hazards. Similarly, Or–And and Nor–Nor design cannot have static-1 hazards.

There! Got rid of half the problem. How about the other half? That's not too bad. As we design a circuit using the K-maps, we must never leave an uncovered pair. I'll illustrate, since this isn't obvious.

Figure 5.5-3 shows a map of a simple three-variable function. If we implement it as we have learned so far, we'd write $W = AC' + A'B'$. That covers two of the pairs. But there's a hazard and it does lead to a glitch. The circuit is shown in Fig. 5.5-4. In its timing diagram (Fig. 5.5-5) we can see a glitch when $B = 0$ and $C = 0$ and A changes from 1 to 0.

Look at the map in Fig. 5.5-3 again. Can you find an uncovered pair? Not if you are looking just at covered 1s, because all the 1s are covered. But there is a pair, the upper-left and upper-right corners, that aren't covered *as a single pair*. A hazard is created by an uncovered pair. That pair is itself uncovered. That is a hazard.

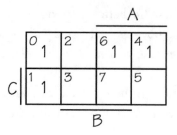

FIGURE 5.5-3: Two pairs?

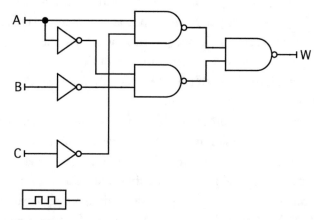

FIGURE 5.5-4: Circuit for W with glitch.

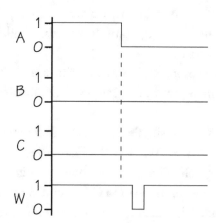

FIGURE 5.5-5: Glitch in W.

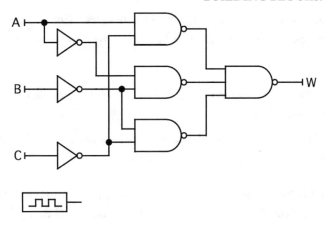

FIGURE 5.5-6:  Glitch-free circuit for W.

Notice that both 1s in the pair have $B = 0$ and $C = 0$. When we changed A from 1 (upper-right corner) to 0 (upper-left corner) we saw a glitch. The uncovered pair is the culprit. It looks like changing A from 0 to 1 should also cause a glitch, but it doesn't. Hazards don't always produce glitches, and when they do, not always in both directions. (The reason here has to do with the inverter delay on the A input.) This static-1 hazard does produce a glitch.

We can eliminate the hazard and the possibility of a glitch by covering the uncovered pair. The term $B' \bullet C'$ will do it, so $W = A \bullet C' + A' \bullet B' + B' \bullet C'$ is the result. The circuit of Fig. 5.5-6 eliminates the glitch but it costs another Nand gate (and a larger output gate).

Suppose we wanted Or–And or Nor–Nor. Is there a hazard problem? The technique, remember, is to design using the 0s and then complement the output. The map in Fig. 5.5-7 shows the 0s. From these the output is $W' = A' \bullet B + A \bullet C$. Complementing the result gives an Or–And form: $W = (A + B') \bullet (A' + C')$.

Is there a hazard here? How about the two 0s in the middle at the bottom? Do they have the same covering? I'll leave the rest of this and whether a glitch results to the problems. This is a static-0 problem.

The rules for avoiding hazards and possible glitches are simple:

- Stay with two-level logic to avoid dynamic hazards.
- In And–Or and Nand–Nand logic, static-0 hazards don't exist. Leave no uncovered adjacent 1s to avoid static-1 hazards.
- In Or–And and Nor–Nor logic, static-1 hazards don't exist. Leave no uncovered adjacent 0s to avoid static-0 hazards.

Must we design to avoid hazards? Only in a few cases. Most of the time hazards, even if they lead to glitches, aren't a problem. There are two situations where they are a problem:

- Feedback sequential circuit design (unclocked) where they can be a big problem. We will not design any of these.

- Output driving an LED. The glitch may cause a flash that will probably bother the user (and won't speak very well of the designer).

Other than these, we can ignore hazards. So when a problem asks you to design a certain circuit, you can safely ignore the hazards. But do stick to two-level design!

## 5.6    DESIGN EXAMPLE

To finish off the combinational circuit part of this book, I'm going to design one circuit. But combinational circuits by themselves aren't terribly interesting. Logic almost always involves time and memory. In other words, we almost always find ourselves designing sequential circuits. But since we aren't that far in the course yet, I'll assume the existence of the sequential part so that we focus on the combinational stuff.

A vending machine needs to count coins put into the coin slot. This vending machine accepts nickels, dimes, and quarters only. We need a circuit that will tally the coins as they come in. What is done with this information is not part of our problem.

This is an example of divide-and-conquer, of modular design. The overall problem would be to design the whole controller for a vending machine. One part of that controller would be the part that controls the actual vending while keeping track of stock. Another would be the part that decides there's enough money and triggers the vending operation. Another part would compute and return change. So our part, which tallies incoming coins, is just a part of a bigger system.

Modular design requires careful specification of the interface between modules. If these are tightly tied down, then the modules should work together when they are all assembled into the complete system. In our example, we presume that this has been done.

Here's what we know:

- Each coin "announces" itself by a signal on an appropriate wire. These signals are pulses that are properly timed for our use.

- The count of the coins is stored in a register. Our circuit is to receive data on the value already stored in the register, add to it the value of the coin just announced, and return the new value to the same register. All of the timing of the register has been done for us.

- The design team has not yet decided on the way to encode the coin count. It has also not yet decided on how large the count can be. They've left that to us because we have to think about the whole coin counting process anyway. We will then propose to the rest of the team what we recommend.

Hmmm, how big to make things? I think two facts will tell us how to make our decision. First, logic devices tend to come in four-bit and eight-bit sizes. That is, the common groupings are 4 and 8. Second, the machine needs to be able to handle a reasonable deposit of money. Pop machines normally take at least 50%. Bottled water runs at least $1. So we need to accept well over $1.

Suppose we limit the count to four bits. That's a maximum of $2^4 - 1 = 15$. That's hardly enough money! Well, how about eight bits, which is $2^8 - 1 = 255$. Hmmm, $2.55 would seem to be plenty—unless inflation gets the best of us very quickly (before our machine is obsolete).

But what advantage is there to counting pennies? After all, the machine doesn't accept pennies. So perhaps we should count nickels, or in other words, in steps of 5%. Then four bits would handle up to 75% . Nope, four bits won't do. Eight bits will handle up to $12.75, which is certainly more than we can foresee.

OK, we propose to the team that we count in steps of 5% and that we build this with eight-bit counters. They argue that we should reduce this to six bits and $3.15 maximum to save money. But we point out that since logic comes in groups of four and eight, using only six would still require eight-bit logic devices. So everyone finally agrees with our proposal.

Now that we have all this settled, we need a list of signals that we are to work with:

- QP, DP, and NP are the pulses from the three coin receivers.
- C7, C6, ..., C0 are the eight register outputs that provide the current value of the coins received.
- D7, D6, ..., D0 are the inputs to the register that holds the coin count. Our circuit is to provide new values of the D signals based on the C signals and the coin just received.
- C8 is the output that indicates that the count has overflowed the maximum of 255 nickels.
- CLK is a clock that runs the whole thing, even though we are not concerned with the sequential circuit itself.

- Reset is a signal that is provided to restore the count to zero after the product has been vended. This signal isn't part of our problem.

Now what? Let's see, basically we have a number in binary (the C signals). We want to increase that number by the value of the coin just received. So if a nickel is received, we add 1 to the count. A dime means we add 2. A quarter means we add 5.

I suppose we could figure out a clever circuit for taking the C signals, adding 1, 2, or 5, and returning the result as the D signals. That would take some thinking and is probably wasteful. After all, we do have devices that do addition for us. Why not use those?

So if we use an adder, which one? Rummaging in catalogs uncovers an eight-bit full adder. That sounds like the device to use. Here are the connections:

- One set of eight inputs will be the current count—the C signals.
- The other set of inputs will be either 1 or 2 or 5, depending on the coin that was just received.
- The output will be sent to the register—the D signals.
- The carry out of the upper end of the addition will provide the overflow signal—C8.
- The carry in will be 0.

How do we provide the 1 or 2 or 5? These values have to be eight-bit numbers. Let's call the eight bits of this input A7, A6, ..., A0. The five high-order bits will be zero: $A7 = A6 = A5 = A4 = A3 = 0$. A2 will be 1 for a quarter only, A1 will be a 1 for a dime only, and A0 will be a 1 for a quarter or a nickel. In other words, the three possible binary eight-bit inputs are 00000001, 00000010, and 00000101.

For the three coin signals QP, DP, and NP, the logic for the adder input is straightforward. $A2 = QP$, $A1 = DP$, and $A0 = QP + NP$.

Gee! That about does it. The circuit that I've designed in words is shown in Fig. 5.6-1. We have done our job. But we still need to integrate this circuit with the coin-receiving logic and the register.

Although we aren't ready to design the rest of the circuit, let's look at it. Figure 5.6-2 shows the complete circuit. The register is the block on the right. Because it is a group of flip-flops, it receives both the clock and the reset signals. The three blocks on the left labeled "One-shot" are used to clean up the coin signals. The coin mechanism can produce the coin signal at any time, depending on when the coil slides through. But this would be unsynchronized with the clock, which creates problems in sequential circuits. So the one-shot circuit performs the synchronization.

Now we have a working coin counter for this vending machine.

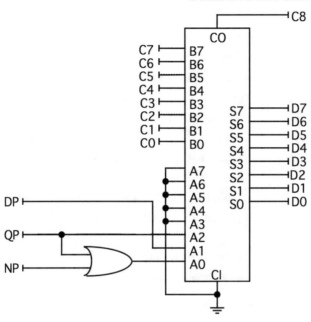

**FIGURE 5.6-1:** Adder for coin counter.

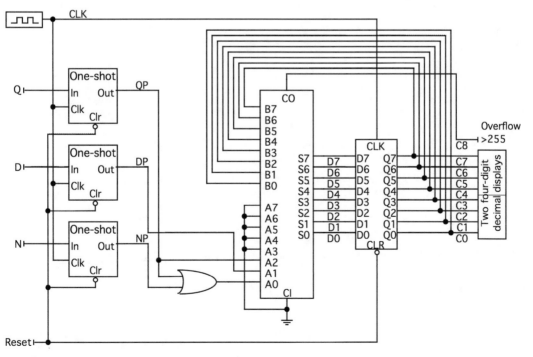

**FIGURE 5.6-2:** Complete coin counter.

## 5.7   SUMMARY

We really have gone a long way in learning about the design of logic circuits. Yet you might be a little concerned about your ability to do a design like the one in the previous section.

Well, the hard parts are not the actual logic development. The hard parts are the thinking that gets us to the actual logic specifications, along with the need for some knowledge of what devices are available and which ones might do the job. Much of this comes with practice.

We've only scratched the surface of "larger blocks" with decoders (minterm generators), multiplexers (for one-chip functions), arithmetic-logic units, and ROMs and PLDs. There are many other types of blocks that can be useful at times. But those we've seen are pretty common.

A study of hazards and resultant glitches shows one problem a designer can confront. We need to design with hazard prevention in mind, because glitches often make trouble when we least expect it.

In the next chapter, we'll take up sequential logic, although we've seen several examples already. But all of our knowledge so far will be important, since the major part of any sequential circuit is combinational.

CHAPTER 6

# Sequential Circuits: Now You Have Time!

Combinational logic doesn't depend on time. With the exception of delays through circuits, time is not a factor at all. The only common problem with time is the glitch caused by those delays. The only place we have seen circuits involving time so far has been in a few places where I have used sequential circuits as examples. Even then, my only real concern was the combinational portion of the circuit.

Sequential logic involves time in an important way: memory. Sequential logic "remembers the past," but not very well. Sequential logic remembers only the most immediate past event, but that's memory nevertheless.

In this chapter I will introduce some concepts that go into our understanding of sequential circuits. Our goal will be to design such logic, but some analysis has to come first.

I'll start by presenting an example of a sequential logic circuit. Then I'll analyze it until we know what it does. Finally, the design process is the reverse of the analysis process. Almost.

But wait, you say, I don't know anything about sequential circuits! Aren't there some theories that I need to know? How about a bunch of definitions? What are these things made out of? It sounds like I'm going to be lost!

No, just stick with me for a while. I'll present the circuit and then we'll study it. Out of this will come not only the analysis process but also the definitions and terms and theories and whatever else we need.

## 6.1 SEQUENTIAL LOGIC ANALYSIS

Start with the circuit of Fig. 6.1-1. It's a sequential-logic circuit. Oh, how do I know? It's tempting to say that the clock gives it away. Not so! It's two other things. First, the circuit involves *memory* in the form of two flip-flops (the rectangular boxes). Second, it's *feedback* from the memory output around to the input.

Hey, we've already got one definition, that of a sequential circuit. Figure 6.1-2 shows a generic version of sequential logic. Compare the parts of this circuit with the elements in my

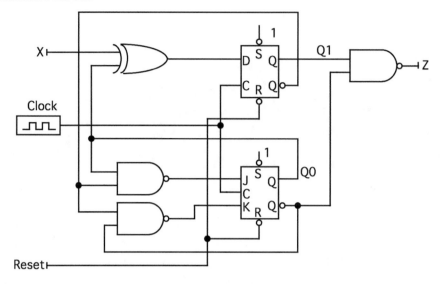

FIGURE 6.1-1: Example for analysis.

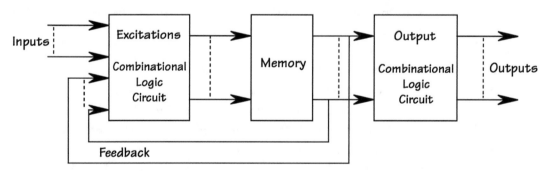

FIGURE 6.1-2: Sequential logic circuit.

example. The gates on the left in the example are the excitation combinational logic circuit. The two flip-flops in the middle are the memory. The gate on the right is the output combinational logic circuit. The feedback comes from the memory and enters the logic on the left. The input is obvious.

Yes, you say, but you've mentioned flip-flops several times without telling me what they are. So let's see if we can define them as something other that rubber footwear.

### 6.1.1   Flip-Flops
A flip-flop is the simplest (logically) memory element. It stores one bit. That bit can be either a zero or a one. We call this stored bit the *state* of the flip-flop.

OK, now for a definition. The *state* of a flip-flop is the bit it is storing. One flip-flop can be in either of two states, zero or one. A second flip-flop can also be in either of two states. So the combination of the two flip-flops has four different possible states: 00, 01, 10, or 11.

Let's look at state a little differently. I presume you are sitting down at the moment, or at least you are not standing. I define your state as either standing or not standing. Have you another person in sight? Is that person standing or not standing? Notice that you can arrange yourself and that other person into four distinct positions.

Imagine now a room full of people. Let your mind have them all sitting. Then the state of the room is all persons not standing. Now one stands up. Aha, a *different state* of the room! Two others now stand—another different state. And so on. If there are ten people in the room, any of whom could be standing or not, the group of people has 1024 different possible states. That's because each one could be standing or not. The possible combinations are $2 \times 2 \times 2 \times 2 \times 2 \times 2 \times 2 \times 2 \times 2 \times 2 = 2^{10} = 1024$.

Ten different flip-flops have 1024 different possible states too. And a hundred flip-flops? $2^{100}$, which is about $1.27 \times 10^{30}$, around a billion. If you want to blow your mind, consider that a computer is a sequential circuit. And how many possible states does it have? If you have a 32-meg memory, hmmm. A "meg" is really $2^{20}$. "32" is $2^5$. That makes 32 megs $= 2^{25}$ bytes. But a byte is eight bits. So your memory is $2^{33}$ bits. That's 8,589,934,592 bits. Each can be in one of two states. So your memory has $2^{8,589,934,592}$ possible states, which is about $10^{2,585,828,010}$, which is a pretty large number.

So it seems that talking about the state of your computer memory doesn't make sense? No, not at all! The memory at any moment has a certain state. We can talk about these. We just can't list all possible states.

OK, so a flip-flop has two possible states. That means our circuit of Fig. 6.1-1 has four possible states. I've labeled the outputs of the flip-flops Q1 and Q0. Since the state of each flip-flop can be either 0 or 1, then Q1 and Q0 are bits. Together they imply a binary number. I'll generally presume that Q0 is the low-order bit of that number. Hence, the four possible states become four possible values of Q1 and Q0, namely, Q1Q0 $=$ 00 or 01 or 10 or 11.

Whoa, you say! You have been carrying on about states for so long that you've forgotten about flip-flops! Well, perhaps, but we do need to get the concept of state straight first. But OK, you win, let's look at flip-flops.

Our example circuit of Fig. 6.1-1 has two different types of flip-flops in it. They are called "D" and "JK" flip-flops, respectively. In addition, each has a clock input, so they are called "clocked" flip-flops as well.

Figure 6.1-3 shows the "transition table" for a D flip-flop. Time is involved through the words "current" and "next." Take a look at the table and notice that the current state has no

| Current state Q | Excitation D | Next state Q |
|---|---|---|
| 0 | 0 | 0 |
| 0 | 1 | 1 |
| 1 | 0 | 0 |
| 1 | 1 | 1 |

FIGURE 6.1-3: Transition table for D flip-flop.

| Current state Q | Excitation J K | Next state Q |
|---|---|---|
| 0 | 0 0 | 0 |
| 0 | 0 1 | 0 |
| 0 | 1 0 | 1 |
| 0 | 1 1 | 1 |
| 1 | 0 0 | 1 |
| 1 | 0 1 | 0 |
| 1 | 1 0 | 1 |
| 1 | 1 1 | 0 |

FIGURE 6.1-4: Transition table for JK flip-flop.

effect on the next state. Only the excitation D influences the next state. The next state Q equals the excitation D.

Now here's a most important point. The value of the excitation D at the *beginning of the clock tick* determines the *next* state of the flip-flop. (To be 100% technical, *beginning* is the rising edge of the clock pulse.)

A second most important point is that we are allowed to be concerned about the state of a flip-flop *only at the beginning of the clock tick*. This allows time for the flip-flop to start changing and have time to finish before we look at the state again.

Oooh, I've just defined a *synchronous* sequential circuit. It's one that operates in synchronism with a clock, where we look at flip-flop outputs only at the clock tick.

That's why I have called the "D" column of Fig. 6.1-3 the *excitation:* the incoming signal excites the flip-flop, causing it to assume a new state (or perhaps to stay in the same state).

Now how about a JK flip-flop? Figure 6.1-4 shows the table for the JK flip-flop. Notice four things:

- If JK = 00, the state of the flip-flop doesn't change.
- If JK = 01, the next state is 0. To say this another way, K is the "reset" input, meaning that it forces the next state to 0 no matter what it was before.

| Current state | Excitation | | Next state |
|---|---|---|---|
| Q | S | R | Q |
| 0 | 0 | 0 | 0 |
| 0 | 0 | 1 | 0 |
| 0 | 1 | 0 | 1 |
| 0 | 1 | 1 | ? |
| 1 | 0 | 0 | 1 |
| 1 | 0 | 1 | 0 |
| 1 | 1 | 0 | 1 |
| 1 | 1 | 1 | ? |

FIGURE 6.1-5: Transition table for SR flip-flop.

| Current state | Excitation | Next state |
|---|---|---|
| Q | T | Q |
| 0 | 0 | 0 |
| 0 | 1 | 1 |
| 1 | 0 | 1 |
| 1 | 1 | 0 |

FIGURE 6.1-6: Transition table for T flip-flop.

- If $JK = 10$, the next state is 1. Hence J is the "set" input, forcing the next state to 1.
- If $JK = 11$, the state of the flip-flop changes. If it was 1 at the clock tick, it will be 0 by the next clock tick, and vice versa.

While we are at it, there are two other fairly common flip-flops, the SR flip-flop and the T flip-flop. Figure 6.1-5 shows the SR flip-flop. S is the "set" input and R is the reset input. Making both of them 1 (i.e., $SR = 11$) confuses the SR flip-flop and the next state is indeterminate. Figure 6.1-6 shows the T flip-flop, which is triggered from one state to the other by the T input.

Well, now that we know about states and flip-flops and synchronous and setting and resetting, can't we get on with the analysis?

## 6.1.2   Tabular Analysis of Example

I'll repeat the example as Fig. 6.1-7 just so we can see it better. (I've added a few labels.) The first step in my analysis is to write the *excitation functions* and the *output functions*. These come directly from looking at the combinational circuits. Don't forget that the combinational circuit

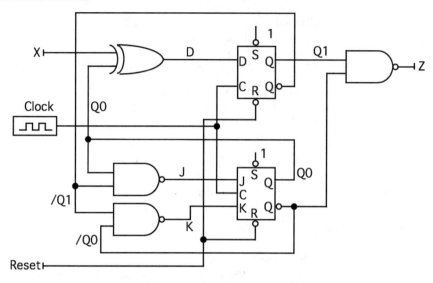

FIGURE 6.1-7: Example for analysis with labels added.

on the left furnishes the excitations that drive the flip-flops and the combinational circuit on the right furnishes the output.

The excitation functions are gotten by analyzing the gates on the left:

$$D = Q_0 \oplus X,$$

$$J = \overline{(\overline{Q_1} \bullet Q_0)} = Q_1 + \overline{Q_0},$$

$$K = \overline{(\overline{Q_1} \bullet \overline{Q_0})} = Q_1 + Q_0.$$

The output function is the gate on the right:

$$Z = \overline{(Q_1 \bullet \overline{Q_0})} = \overline{Q_1} + Q_0.$$

Now the excitations are easy to tabulate. Figure 6.1-8 shows the truth-table form of the excitation functions. This table is called the *excitation table*, which is a clever name, don't you think? It tells us that, for a given "current state" Q1Q0 and a given input X, the flip-flops will be excited by certain values.

Notice especially the labels on the columns in this table. The words are important: "Current Q1 Q0," "Input X," "Excitation D," and "Excitation JK." Make it a habit, please, to label such tables *completely* to help avoid mistakes in both tabulation and understanding.

Now we need to tabulate what happens to the flip-flops because of these excitations. In Fig. 6.1-9 I've added the "Next Q1 Q0" columns. We find these next states through our new knowledge of how the D and JK flip-flops work:

| Current Q1 Q0 | | Input X | Excitation D : J K | | |
|---|---|---|---|---|---|
| 0 | 0 | 0 | 0 : 1 | 0 |
| 0 | 0 | 1 | 1 : 1 | 0 |
| 0 | 1 | 0 | 1 : 0 | 1 |
| 0 | 1 | 1 | 0 : 0 | 1 |
| 1 | 0 | 0 | 0 : 1 | 1 |
| 1 | 0 | 1 | 1 : 1 | 1 |
| 1 | 1 | 0 | 1 : 1 | 1 |
| 1 | 1 | 1 | 0 : 1 | 1 |

FIGURE 6.1-8: Excitation table.

| Current Q1 Q0 | | Input X | Excitation D : J K | | | Next Q1 Q0 | |
|---|---|---|---|---|---|---|---|
| 0 | 0 | 0 | 0 : 1 | 0 | 0 | 1 |
| 0 | 0 | 1 | 1 : 1 | 0 | 1 | 1 |
| 0 | 1 | 0 | 1 : 0 | 1 | 1 | 0 |
| 0 | 1 | 1 | 0 : 0 | 1 | 0 | 0 |
| 1 | 0 | 0 | 0 : 1 | 1 | 0 | 1 |
| 1 | 0 | 1 | 1 : 1 | 1 | 1 | 1 |
| 1 | 1 | 0 | 1 : 1 | 1 | 1 | 0 |
| 1 | 1 | 1 | 0 : 1 | 1 | 0 | 0 |

FIGURE 6.1-9: Excitation/transition table.

- For the D flip-flop, the next state is the same as the current D input, so since Q1 is a D flip-flop, I have just copied the D column into the Q1 column.

- For the JK flip-flop, I use the set–reset idea. In the first and second rows, J = 1 and K = 0, so the flip-flop is to be set, which yields a next Q0 of 1. In the next two rows, J = 0 and K = 1, so the next Q0 is 0 ("reset"). In the last four rows, J = K = 1, so the flip-flop state changes.

Finally, in Fig. 6.1-10 I've added the output Z. The result is the *excitation/transition/output table* for my circuit. These all neatly combine into one table.

Gee, aren't those tables nice? But what is this circuit doing? There's another way to see what is happening, this one more in the sense of a flow chart. Figure 6.1-11 shows the *state diagram* for my circuit. It shows the states, the outputs, and what moves the circuit from one state to another.

| Current Q1 Q0 | Input X | Excitation D ¦ J K | Next Q1 Q0 | Output Z |
|---|---|---|---|---|
| 0  0 | 0 | 0 ¦ 1  0 | 0  1 | 1 |
| 0  0 | 1 | 1 ¦ 1  0 | 1  1 | 1 |
| 0  1 | 0 | 1 ¦ 0  1 | 1  0 | 1 |
| 0  1 | 1 | 0 ¦ 0  1 | 0  0 | 1 |
| 1  0 | 0 | 0 ¦ 1  1 | 0  1 | 0 |
| 1  0 | 1 | 1 ¦ 1  1 | 1  1 | 0 |
| 1  1 | 0 | 1 ¦ 1  1 | 1  0 | 1 |
| 1  1 | 1 | 0 ¦ 1  1 | 0  0 | 1 |

FIGURE 6.1-10:  Excitation/transition/output table.

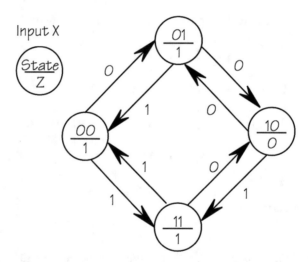

FIGURE 6.1-11:  State diagram.

Notice the labels in the upper-left corner. These tell the reader what is being shown. The input is named X, which are the numbers written along the arrows. The circle contains two items, the binary value of the state of the circuit and the output for that state.

Let's read in English a bit of the diagram. Start with the circle with the 00 inside, state 00. I can read that circle as, "If the state is 00, the output is 1."

The transitions are given by the arrows. The arrow labeled 0 leaving the 00 circle is read, "If the state is 00 and the input is 0, the next state is 01." The other arrow leaving state 00 is read, "If the state is 00 and the input is 1, the next state is 11." And so on.

The state diagram helps us see the pattern of changes as the inputs change. But don't lose sight of the fact that *each transition happens on a clock tick.* The start of a clock tick starts movement along the appropriate arrow, arriving at the next state before the beginning of the next clock tick. This transition from one state to the next takes time.

### 6.1.3   Analysis Steps
It's useful to list the steps we've gone through:

1.  From the combinational circuits, write the excitation functions and the output functions.
2.  Create the excitation table from the functions that drive the flip-flops.
3.  Write the next states of the flip-flops from the tabulated excitations and your knowledge of how flip-flops respond. This gives the transition table.
4.  Add the output table from the output functions.
5.  Draw the state diagram with proper labels.

Well, what's next? First, a few additional pieces of information that you should know to be literate. Then we'll do some designing.

## 6.2   LATCHES AND FLIP-FLOPS
Flip-flops include a clock input, as we have just seen. The flip-flop doesn't do anything until the clock ticks. And even when the clock ticks, the only time something happens is on the leading edge of the tick. Figure 6.2-1 shows a clock and defines the various edges.

We can look at this idea of "something happens" in a slightly different way. We can say that the flip-flop ignores its inputs except when they are *sampled* by the clock. For the flip-flops that we've seen, this sampling takes place on the leading edge of the clock pulse.

Why is this important? Consider what is happening in time. Suppose the flip-flop "looks" at its input all the time, not just at the clock edge. Suppose that input causes the bit stored in the flip-flop to change. That change appears at the output. That output typically is fed back to the input in some way. That change changes the input. The flip-flop is still looking, so it changes again. In other words, we have unintended behavior.

Sampling the inputs means that changes caused by those inputs don't come back around and make new changes before those changes are wanted. How this sampling is done isn't going to be important to us here.

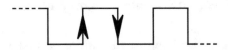

FIGURE 6.2-1: Clock showing rising & falling edges.

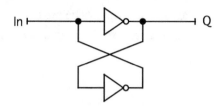

**FIGURE 6.2-2:** Inverter latch.

Well, this all implies that there must be other kinds of flip-flops out there, ones that don't sample their inputs, right? Yup! But we don't call them flip-flops. That term is reserved for the kind that sample. We'll call the other ones *latches*.

Latches are very simple. They can be made from just two gates. All we need are two two-input gates, Ands or Ors, that have complemented outputs. In other words, we can make a latch from two Nands or two Nors. I'll do that in a moment.

### 6.2.1   Inverter Latch

But a latch can be made from just two inverters, although we have to do something slightly immoral, illegal, and fattening. Figure 6.2-2 shows a latch using two inverters. Let's see how it works:

- Suppose In = 0. Then Q = 1. That 1 is the input to the lower inverter. The output of the lower inverter is 0, which feeds back to In. But since In is already 0, everything is stable.

- Now suppose we make In = 1. Keep in mind that gates have delays, so nothing happens instantaneously. The new input passes through the inverter, making Q = 0. That change passes through the second inverter, making its output 1. That matches the new value of In, so everything is stable again.

Be sure that you see how important time is. In this circuit, the delay through the inverters is what allows us to change the setting of the latch. We change In and the change wanders through two inverters, finally hooking up with In and settling down.

So this looks like a pretty nice circuit. Uhuh! There's a flaw, or at least something not quite right. Have you caught it? How do I change In? Sure, it's an input. But look at the wiring. Isn't In directly wired to the output of the lower gate? When we change In, aren't we forcing that output to a different value? And that output is already being held electrically by the inverter to a different value.

So are we creating lightning? Suppose the output of the lower inverter is a 1 and is 4.5 volts (a usual level for gates with 5-volt power supplies). Now apply a 0 to In, which means

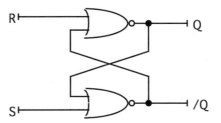

FIGURE 6.2-3: SR latch.

grounding In. That grounds 4.5 volts, which should create a problem. Fortunately, the inverter's output is somewhat forgiving and allows this to happen without complaining. So the latch does work. But it's not the best circuit in the world.

## 6.2.2  Better Latch

We can do better by using a "two-input inverter," something that doesn't exist. Such gates exist, though, in the form of Nands and Nors. I could choose either, but I will choose a pair of Nors as shown in Fig. 6.2-3. Let's see how this circuit works:

- Suppose R and S are both 0 and that $Q = 0$. That means that /Q had better be 1 if my notation is to mean anything. Let's see. Follow $Q = 0$ through the circuit. The inputs to the lower Nor are both 0. The output is therefore 0', which is 1. So far so good. That 1 makes the inputs to the upper Nor gate 1 and 0. Its output is therefore 1' = 0. There! the loop is completed and the latch is stable.

- Now make S = 1. The inputs to the lower gate are now 1 and 0, so its output is 1' = 0. The inputs to the upper gate are now both 0, so its output is 0' = 1. Hence $Q = 1$. Follow that 1 back to the lower gate, where the inputs are now both 1. Its output is 1' = 0, which is where we came in. The latch is stable.

- Now make S = 0 again. The inputs to the lower gate change from 1 and 1 to 1 and 0. But this doesn't change the output of the lower gate, so the latch remains stable.

We could go on for quite a while here, but I won't. What has happened is that the S input has *set* the latch, changing it from 0 to 1. If you now track through what happens when you make R = 1 and then R = 0, you'll find that you have *reset* the latch, changing it from 1 to 0.

This contraption is called an *SR latch* because you can set and reset it. *Set* always means *to 1* and *reset* always means *to 0*. It's a latch because there is no sampling of the inputs and there is no clock. The circuit responds any time you change an input.

| Current state | Excitation | | Next state |
|:-:|:-:|:-:|:-:|
| Q | S | R | Q |
| 0 | 0 | 0 | 0 |
| 0 | 0 | 1 | 0 |
| 0 | 1 | 0 | 1 |
| 0 | 1 | 1 | ? |
| 1 | 0 | 0 | 1 |
| 1 | 0 | 1 | 0 |
| 1 | 1 | 0 | 1 |
| 1 | 1 | 1 | ? |

FIGURE 6.2-4:  SR latch transition table.

| Excitation | | Next state |
|:-:|:-:|:-:|
| S | R | Q+ |
| 0 | 0 | Q |
| 0 | 1 | 0 |
| 1 | 0 | 1 |
| 1 | 1 | ? |

FIGURE 6.2-5:  SR latch: simpler table.

The transition table for the SR latch is shown in Fig. 6.2-4. Notice what happens when both S and R are 1. The circuit is unstable and unpredictable. So if we implement something using an SR latch, we must make sure that we never make both inputs 1 at the same time.

Figure 6.2-5 shows a slightly different and shorter way of saying the same thing. The next-state column is given in terms of the present state. The current state is named Q and the next state is named Q+. The first row says that if $S = R = 0$, the next state is the same as the current state, and so on.

## 6.2.3   Debouncing a Switch

Before I quit talking about latches, here's one application of an SR latch—debouncing a mechanical switch. A switch, any switch, is designed to convert mechanical motion into electrical information by moving contacts. But any mechanical system has some dynamic properties that sometimes get in our way. A switch contact is something metallic that moves toward or away from another metallic something. These somethings have mass, so they don't move instantaneously. Once they contact one another, they possibly rebound, thereby losing contact.

All switches have bounce in the contacts, some worse than others. I have seen some switches that sometimes bounced six times, making and breaking the contact each time. A

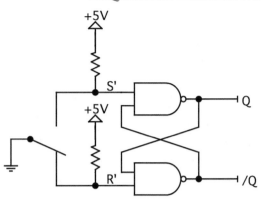

FIGURE 6.2-6: Debouncing a switch.

poorly designed logic circuit might interpret this as a string of distinct inputs rather than the single intended one.

We need a circuit that will take all these bounces and interpret them as one switch movement. The circuit of Fig. 6.2-6 does this. I am using a variation of the SR latch, sometimes called the S'R' latch because its actions are reversed. It uses Nand rather than Nor gates.

Suppose the switch is as shown in the lower position. The lower Nand gate has at least one of its inputs grounded, which is 0. Hence, the output of the lower Nand is 0' = 1. Both inputs of the upper Nand are 1, so its output is 1' = 0. The result is that Q = 0.

Now move the switch to the upper contact. That gives the upper Nand a 0 on one input, so its output becomes 0' = 1. Meanwhile, the switch has changed the lower Nand's input to 1 because the pull-up resistor pulls the line up to 5 volts. Since Q now is 1, this lower Nand has a 1 on both inputs, so its output is 1' = 0. That feeds back to the upper Nand .... The circuit is stable.

Suppose the switch now bounces, leaving the upper contact. It isn't going to touch the lower contact because the mechanical distance is too great. (A switch that does that should probably be tossed out!) The upper Nand sees a 1 on the switch input because of the pull-up resistor. But the other input is 0, so the output of the Nand is still 0' = 1. The bounce doesn't change the latch.

## 6.2.4   Altogether

It might be useful to have the characteristics of all the flip-flops and latches together in one place before we go on. Figure 6.2-7 shows all the tables. I've used the "current-next" notation of Q and Q+ as I did for the SR latch. Notice that you cannot tell a latch from a flip-flop by looking at these transition tables.

| D | Q+ |
|---|---|
| 0 | 0 |
| 1 | 1 |

| T | Q+ |
|---|---|
| 0 | Q |
| 1 | Q' |

| S | R | Q+ |
|---|---|---|
| 0 | 0 | Q |
| 0 | 1 | 0 |
| 1 | 0 | 1 |
| 1 | 1 | ? |

| J | K | Q+ |
|---|---|---|
| 0 | 0 | Q |
| 0 | 1 | 0 |
| 1 | 0 | 1 |
| 1 | 1 | Q' |

FIGURE 6.2-7: Tables for flip-flops.

## 6.3    DESIGN ELEMENTS

Sequential logic design is pretty straightforward. Well, almost! There are several places where the design process requires something like a crystal ball because the process is not algorithmic. In other words, there are some places where there are not clear rules telling you exactly what to do.

But that doesn't stop us! So let's see what the elements of the design process are:

1. *Word problem*: Start with a word problem that tells what our logic is to do. Word descriptions can be imprecise and confusing, so we must be very careful that we understand what is wanted.

2. *State diagram*: Translate the word problem into a state diagram that shows in graphical form the states and how the transitions are to be made. Also show the outputs. This step is sometimes skipped, going instead directly to step 3.

3. *State table/output table*: Create a table that gives the next states and the outputs for every input transition. At this point, the states usually still have names rather than binary values.

4. *Minimization*: Reduce the number of states if possible. Sometimes we translate the word problem and introduce too many states. If we can reduce the number, our logic is often simpler. An example is a problem where we decided we needed ten states. That requires four flip-flop: three is too few because $2^3 = 8$. If we can reduce the number of states by at least two, we cut out one flip-flop.

5. *Assign state variables*: Assign binary values to each of the named states if they don't have such values already. This step is not algorithmic! For example, suppose there are six states in the problem. There are eight different binary values I can assign to the first state (000, 001, 010, and so on). Then there are seven for the second state. There are

$8 \times 7 \times 6 \times 5 \times 4 \times 3 = 20,160$ different combinations! There are a few heuristics, though, which are rules of thumb for making assignments that help keep the circuit simpler.

6.  *Choose flip-flops*: Somebody has to decide whether to design with JK or D or whatever flip-flops. Sometimes this is already decided: "Use JKs," says the boss. Today, the common choice is D, since that's what popular programmable logic devices provide. This step isn't algorithmic, either. One choice of flip-flop type might give a simpler circuit than another. The only way you can find out is to try them all!

7.  *Excitation table*: Now that you know the flip-flop type, create the excitation table that tells how to get from one state to another.

8.  *Excitation and output functions*: From the excitation and output tables, write the logic functions for the excitation and output combinational logic.

9.  *Logic diagram*: Put the flip-flops, the excitation logic, and the output logic together into a complete logic circuit.

10. *Test it*! Test your design by analysis and by simulation if possible. Make sure it does what you set out to do in step 1.

There—ten steps to getting a sequential circuit design is done. I'll probably avoid doing step 4 by being careful as I think my way through the states needed for the problem. I'll probably stick with D or possibly JK flip-flops in step 6. SR is never simpler than JK. T can be used if J = K for each JK flip-flop.

Enough talking! On with a design or two.

## 6.4   DESIGN EXAMPLES

I'll start with a simple design example and end with a more complicated one. By the time you are done with these, you should be able to do your own with not too much trouble. And I think you'll find that the hardest part is getting from the word problem to the state diagram or state table.

### 6.4.1   Example I

My first example is a classic textbook type of problem that isn't terribly exciting. We always have to do something like this when we can't think of anything practical. It starts with a bit stream that is synchronized to our common click. This is a string of zeros and ones coming on an input line, one at each tick of the clock. We are to design a sequential logic circuit that will detect a particular pattern in the incoming bits.

In this example, the bit stream is to be searched for exactly two consecutive 1s. If such a pair is found, the output is to be 1 for one clock tick. Let's follow the design steps.

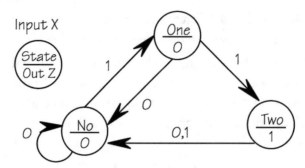

**FIGURE 6.4-1:** Example I: state diagram.

| Current state | Input X | Next state | Output Z |
|---|---|---|---|
| No | 0 | No | 0 |
| No | 1 | One | 0 |
| One | 0 | No | 0 |
| One | 1 | Two | 0 |
| Two | 0 | No | 1 |
| Two | 1 | No | 1 |

**FIGURE 6.4-2:** State/output table.

1. We already have a word description, but we need to add some labels. Call the input bit stream X and the output Z.

2. I'll translate this into the *state diagram* of Fig. 6.4-1. How did I do this?

   First, I defined a starting state, which I named "No" to mean that nothing useful has been detected. There are two possible exits, X = 0 and X = 1. For X = 0, I will stay in the No state because I am looking for 1s. For X = 1, I go to a state that I name "One," meaning I have one 1 that could lead to another 1 and hence a pair.

   This new state has two exits. For X = 0 I am back to where I started, namely, No. For X = 1, I go to a state named "Two" because the stream now has presented a consecutive pair of 1s.

   Finally, this state has two exits, which both lead to the starting state to start look for a new pair.

   The outputs from these states are included in the circles. The output Z = 1 only for the state named "Two" because that's when we know we have a consecutive pair of 1s. Notice the diagram's labeling. I have told what the inputs are and what the notation inside the circle tells.

| Current Q1 Q0 | Input X | Next Q1 Q0 | Output Z |
|:---:|:---:|:---:|:---:|
| 0  0 | 0 | 0  0 | 0 |
| 0  0 | 1 | 0  1 | 0 |
| 0  1 | 0 | 0  0 | 0 |
| 0  1 | 1 | 1  0 | 0 |
| 1  0 | 0 | 0  0 | 1 |
| 1  0 | 1 | 0  0 | 1 |
| 1  1 | 0 | x  x | x |
| 1  1 | 1 | x  x | x |

FIGURE 6.4-3: Example I: transition table.

3. The *State/Output Table* is shown in Fig. 6.4-2. This is merely a translation of the state diagram into table form. Be sure to put column labels in your tables much like those I have used. Your goal is to write these tables so that they make sense. After all, you'll need to design from them and you don't need to confuse yourself!

4. Can we reduce the number of states? There are three, which requires two flip-flops. If we could get this to two, we'd need only one flip-flop. But the names of these states make it pretty clear they can't be combined. "No" means we don't have a 1 at the moment. "One" means we have one of them. "Two" means we have what we are looking for. These all appear to be unique. I guess we are stuck with two flip-flops.

5. Assign *state variables*. I need to pick three binary numbers for the states. There are 24 different ways to do this, but I'll keep this simple. I choose as my starting point "No" = 00. Why? Because it seems obvious, I guess. Also, many flip-flops can be reset through a separate terminal to 0. Hence, we can easily get our circuit to the starting point when the power is turned on. Then I will just count: "One" = 01 and "Two" = 10. The resulting state/output table is shown in Fig. 6.4-3. (State 11 can never occur.)

6. OK, what kind of flip-flop? I am going to start with D flip-flops because they are the simplest to do—and to describe to you. Recall that, for a D flip-flop, the desired next state is the same as the current excitation.

7. For D flip-flops the *excitation table* is the same as the state table, which is shown in Fig. 6.4-3.

8. The *excitation functions* and the *output function* are found from the table of Fig. 6.4-3 just as we have been doing for combinational circuits in the previous chapters. The

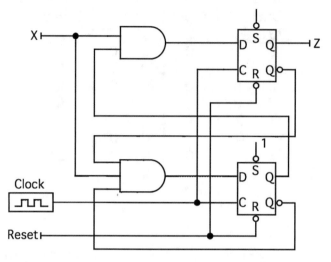

FIGURE 6.4-4:  Example I: circuit using D flip-flops.

results are

$$D_1 = Q_0 \bullet X,$$

$$D_0 = \overline{Q_1} \bullet \overline{Q_0} \bullet X,$$

$$Z = Q_1.$$

9.  The final circuit is shown in Fig. 6.4-4. The excitation functions provide the D inputs to the flip-flops and the output function provides the output Z.

10.  I simulated this and it appears to work.

In step 6 I chose D flip-flops. Could a different choice have produced a simpler circuit? There is no way to know other than to try it. So I am going to redo the design using JK flip-flops and see what happens.

6.  This time choose JK flip-flops.

7.  Creating the excitation table is a little harder now because JK flip-flops have two inputs each. The new table is shown in Fig. 6.4-5, but I need to tell you an easy way to create each of the entries.

There are only four possible combinations of J and K: 0x, x0, 1x, and x1. I use a mantra of sorts to get these into my table. It goes like this (the mantra is in quotes):

• "Now 0, stay 0; don't set, could reset." Why this? To keep a JK flip-flop at 0, we must not set it, so J = 0. But if J = 0, it makes no difference if K = 0 or K = 1,

| Current Q1 Q0 | | Input X | Next Q1 Q0 | | Excitation J1 K1   J0 K0 | | | | Output Z |
|---|---|---|---|---|---|---|---|---|---|
| 0 | 0 | 0 | 0 | 0 | 0 x | | 0 x | | 0 |
| 0 | 0 | 1 | 0 | 1 | 0 x | | 1 x | | 0 |
| 0 | 1 | 0 | 0 | 0 | 0 x | | x 1 | | 0 |
| 0 | 1 | 1 | 1 | 0 | 1 x | | x 1 | | 0 |
| 1 | 0 | 0 | 0 | 0 | x 1 | | 0 x | | 1 |
| 1 | 0 | 1 | 0 | 0 | x 1 | | 0 x | | 1 |
| 1 | 1 | 0 | x | x | x x | | x x | | x |
| 1 | 1 | 1 | x | x | x x | | x x | | x |

FIGURE 6.4-5: Example I: JK excitation table.

because either will leave the flip-flop at 0. So a don't-care is appropriate. In this case $JK = 0x$.

- "Now 0, become 1; set, could reset." If we make $JK = 1x$, $J = 1$ sets the flip-flop. But whether $K = 0$ or $K = 1$, the flip-flop is still going to be set. Recall that $JK = 11$ *changes* the state of the flip-flop, which works here.
- "Now 1, become 0; reset, could set." Here $JK = x1$ because $K = 1$ resets and $J$ can be a don't-care.
- "Now 1, stay 1; don't reset, could set." Hence $JK = x0$, where $K = 0$ prevents resetting.

That's how I constructed the columns of the new excitation table in Fig. 6.4-5. I urge you to use something like my mantra to do this. You are more likely to do the job correctly. Besides, the job is mindless anyhow!

8.  From this table I get the excitation and output functions:

$$J_1 = Q_0 \bullet X, \, K_1 = 1,$$

$$J_0 = \overline{Q_1} \bullet X, \, K_0 = 1,$$

$$Z = Q_1.$$

9.  The circuit is shown in Fig. 6.4-6. Is it simpler than the D implementation? You be the judge.

10. Simulation shows this works just fine too.

## 6.4.2 Example II

A shop-lifting alarm in the drug store has a sensor that catches items that try to go out the door without being purchased. The clerk must be able to reset this alarm. A sensor S becomes 1 if

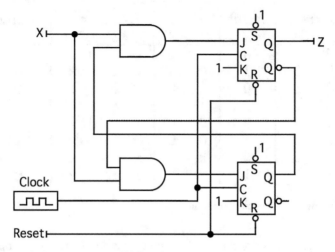

FIGURE 6.4-6: Example I: circuit using JK flip-flops.

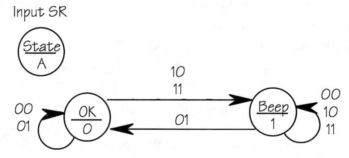

FIGURE 6.4-7: Example II: state diagram.

such an item attempts to escape. The alarm A is sounds. The clerk has a switch R that resets the alarm, but only after the offending item retreats into the store.

Let's go through the design steps, this time with fewer words:

1.  $S = 1$ makes $A = 1$. $R = 1$ makes $A = 0$ if S is now 0.

2.  The state diagram of Fig. 6.4-7 is pretty simple. Two inputs go together on each path. The state names are meaningful. There are labels to tell what the symbols mean.

3.  I assign 0 as the starting state and 1 as the other state before writing the state/output table because the assignment is so simple this time. Hence, Fig. 6.4-8 shows the table with the state variables already assigned.

4.  The number of states cannot be reduced.

5.  I already assigned the state variables in step 3.

| Current | Input | | Next | Output |
| Q | S | R | Q | A |
| --- | --- | --- | --- | --- |
| 0 | 0 | 0 | 0 | 0 |
| 0 | 0 | 1 | 0 | 0 |
| 0 | 1 | 0 | 1 | 0 |
| 0 | 1 | 1 | 1 | 0 |
| 1 | 0 | 0 | 1 | 1 |
| 1 | 0 | 1 | 0 | 1 |
| 1 | 1 | 0 | 1 | 1 |
| 1 | 1 | 1 | 1 | 1 |

FIGURE 6.4-8: Example II.

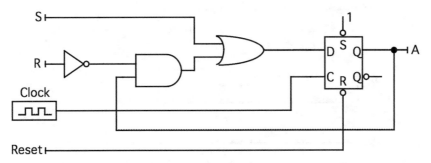

FIGURE 6.4-9: Example II: circuit using D flip-flop.

6. Choose D because I say so.

7. The state table is the excitation table because $D$ = next state.

8. The excitation and output functions are

$$D = S + Q \bullet \overline{R},$$

$$A = Q.$$

9. Figure 6.4-9 shows the circuit.

10. I tested it; it works.

How about a JK solution?

6. Choose JK.

7. The excitations require my mantra again; the result is in the table of Fig. 6.4-10.

| Current | Input | | Next | Excitation | | Output |
|---------|-------|---|------|-----------|---|--------|
| Q | S | R | Q | J | K | A |
| 0 | 0 | 0 | 0 | 0 | x | 0 |
| 0 | 0 | 1 | 0 | 0 | x | 0 |
| 0 | 1 | 0 | 1 | 1 | x | 0 |
| 0 | 1 | 1 | 1 | 1 | x | 0 |
| 1 | 0 | 0 | 1 | x | 0 | 1 |
| 1 | 0 | 1 | 0 | x | 1 | 1 |
| 1 | 1 | 0 | 1 | x | 0 | 1 |
| 1 | 1 | 1 | 1 | x | 0 | 1 |

FIGURE 6.4-10: Example II: JK flip-flops.

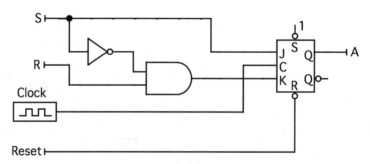

FIGURE 6.4-11: Example II: circuit using JK flip-flop.

8.  The excitation and output functions are

$$J = S, \quad K = \overline{S} \bullet R,$$
$$A = Q.$$

9.  The circuit is shown in Fig. 6.4-11. Is it simpler?

10.  Simpler or not, it works!

### 6.4.3   Example III

This example is going to be more complicated than the first two. Not only does it have lots of possibilities but it will stretch our ability to minimize combinational logic.

The diagram of Fig. 6.4-12 is the crossing of the Whitewater Valley Railroad with Veterans Parkway in Connersville, Indiana. This crossing has a flashing signal, activated by the passage of the train. The track is broken up by insulated joints into three sections. I have called them the North (N), Middle (M), and South (S) sections.

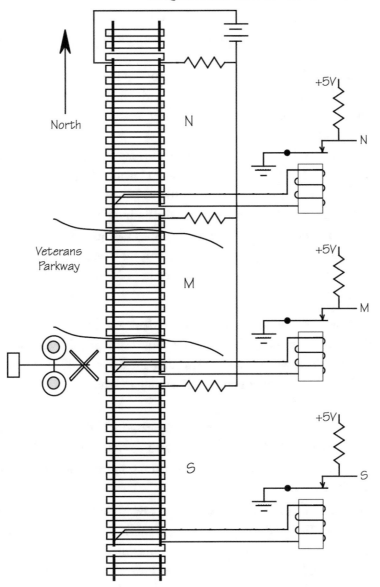

FIGURE 6.4-12:  WVRR highway crossing.

A battery capable of delivering fairly high current is connected to the track. The common lead of the battery is connected to the west rail. The other lead is connected to each section through a resistor. These connections are at one end of each section.

A relay is connected to the rails at the other end of each section. This relay is nothing more than a switch that is activated by an electromagnet. If no current is passing through the coil of the relay, the contact is up, grounding the output wire. When current flows in the coil, the contact moves toward the coil, ungrounding the output wire. The pull-up resistor pulls the output wire up to 5 volts.

Each relay is normally energized, so the contact is pulled down and the output wire is 5 volts, which we'll call 1. When a train enters the north end of the N section, the wheels and axles short across the rails. This prevents current from reaching the relay. So the relay contact releases, making the output 0.

The train progresses through the sections. If it is long enough, the three outputs go from NMS = 111 to 011 to 001 to 000 to 100 to 110 to 111 again. A short train such as an engine alone could cause NMS = 111 to 011 to 001 to 101 to 100 to 110 to 111.

We are to design a circuit that will activate the flashing highway signals. Call the signals F. When a train enters either N or S, the signal is to start flashing (F = 1). The signal is to stop when the train is out of the crossing itself (M). This means that a train entering at N starts the signal. The signal keeps going until the train has left both N and M, but it can still be in S.

Design the circuit! Follow the design steps:

1.  We already have the word problem. Notice there are some holes in it. What happens when the train enters at M, for example? Can this happen? Well, only if it can jump over N or S, which would be an interesting trick! Or could this be caused by some kind of failure? Do we worry about it? I'm going to ignore such things and see how the results come out.

2.  My first attempt at the state diagram is in Fig. 6.4-13. Ugh! It has seven states and three inputs. That's three flip-flops. That means three state variables. Three state variables and three inputs say we will have a six-variable combinational-circuit problem. Maybe I can simplify this.

    I succeeded in reducing the number of states by noticing that some states are almost identical. States "In MN" and "In MS" have much the same actions. The only differences are their entries, which represent the direction of the train. So I tried again. The result is shown in Fig. 6.4-14. Much better! Five states! Ugh! Still three flip-flops.

    Then I realized that states "M–N" and "M–S" are the same so I combined them. ("Same" here means that they have identical outputs F and also the same next states.) Figure 6.4-15 shows a four-state result. Does it work? You find out by trying the various possible input sequences and seeing if the output is always correct.

3.  I did not create a state/output table here. Instead, I assigned the states as shown in the transition/output table in Fig. 6.4-16. This table is in somewhat different form. Basically, I have folded it up. Current states are down the left side; inputs are across

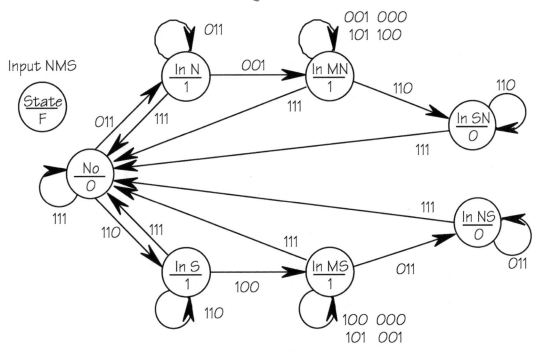

FIGURE 6.4-13: Example III: state diagram, first try.

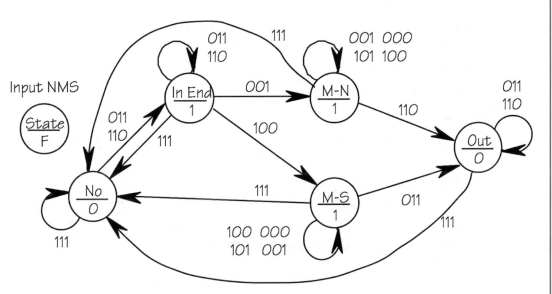

FIGURE 6.4-14: Example III: state diagram, second try.

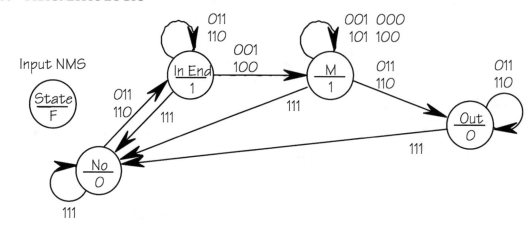

FIGURE 6.4-15: Example III: state diagram, third try.

|  | Current | Next state for input NMS | | | | | | | | Output |
| State | Q1 Q0 | 000 | 001 | 010 | 011 | 100 | 101 | 110 | 111 | F |
|---|---|---|---|---|---|---|---|---|---|---|
| No | 0  0 | xx | xx | xx | 01 | xx | xx | 01 | 00 | 0 |
| In End | 0  1 | xx | 11 | xx | 01 | 11 | xx | 01 | 00 | 1 |
| Out | 1  0 | xx | xx | xx | 10 | xx | xx | 10 | 00 | 0 |
| M | 1  1 | 11 | 11 | xx | 10 | 11 | 11 | 10 | xx | 1 |

FIGURE 6.4-16: Example III: transition/output table.

the top; next states are in the body of the table. For example, the table says that if the current state is 11 and the input is 000, the next state is 11. All the don't-cares come from input situations that are not supposed to arise.

4. My minimization is already done.

5. So is the state assignment.

6. I choose D because they are simplest to use here. I already have enough things to deal with without doubling the number of excitation functions.

7. The excitation table for D flip-flops is the transition table.

8. The excitation and output functions can be gotten by carefully examining the state/output table and thinking about don't-cares. There are so many of them that spotting simple combinations isn't too hard. The results are

$$D_1 = Q_1 \bullet \overline{N} + \overline{M} + Q_1 \bullet \overline{S},$$

$$D_0 = \overline{Q_1} \bullet \overline{N} + \overline{M} + \overline{Q_1} \bullet \overline{S},$$

$$F = Q_0.$$

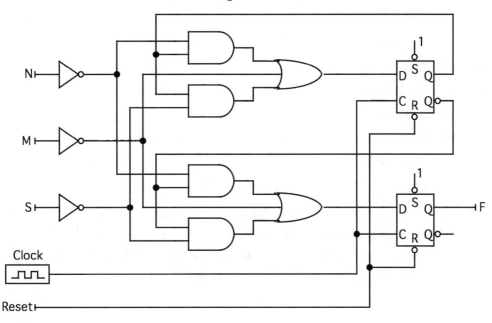

FIGURE 6.4-17: Example III: crossing-signal circuit.

9. The circuit is shown in Fig. 6.4-17.

10. Simulation says the circuit works. But how does it handle all the don't-care situations? An important question in railroad safety is whether the circuit is fail-safe. What happens if a portion of the track circuit fails? What happens if the train somehow appears in M first? I'll leave these questions to the homework. (Yes, I know, grrrrr!)

That's a pretty complicated design problem, but the steps tell us how to go through it rather easily. The hardest parts are getting the number of states down and then dealing with more than four variables.

## 6.5   SUMMARY

Sequential circuits involve time and memory. What we've seen here is memory in the form of flip-flops. Other forms are possible, but we'll stick with what we have. D and JK flip-flops are the most versatile and common ones.

An analysis example gave us a series of steps to analyze a sequential circuit. Reversing the order of these and introducing some non-algorithmic decisions yielded the ten design steps:

1. Word problem.
2. State diagram.

3.   State/output table.

4.   State minimization.

5.   State variable assignment (binary values).

6.   Flip-flop choice.

7.   Excitation table.

8.   Excitation and output functions.

9.   Logic diagram.

10.  Testing.

Finally, we've done some design, which is really the only way to learn what to do. The last example, which appears to be a simple system, turned out to be rather large.

Where from here? Just as we had building blocks for combinational logic, so too we have them for sequential logic. That's what we'll do in the next chapter.

CHAPTER 7

# Counters and Registers: More Building Blocks

Ages ago, at the beginning of Chapter 5, I said that we didn't often design at the very lowest level, that there were building blocks that did some of the "larger" common functions for us. We added decoders and multiplexers, arithmetic-logic units, and memory to our collection of parts.

Sequential logic is no different. So far, we've been working with individual flip-flops and their excitation and output functions. There are common combinations that help us here, devices such as counters and shift registers. In this chapter, we are going to see how some of these can be applied.

There's also another form, or perhaps variant, of sequential logic that is often useful. We have seen one way of doing output, a form that is called a *Moore machine*. In this form, our outputs have been directly associated to states. In other words, the outputs have been in the circles of our state diagrams. Another form, the *Mealy machine*, associates the outputs with the paths on the state diagram. This isn't a bad deal sometimes.

## 7.1 COUNTERS

Counters are nothing more than logic circuits that count. That doesn't seem too difficult! And there can't be that many ways of counting. But that isn't true—there are lots of ways to count. These counter chips do only a few of them, though, the most common ones.

How could we possibly count? 1-2-3-4. . .! In digital logic, though, our count is limited by the size of the counter. It's limited by the number of bits the counter can hold. And some counters count only to ten (really 0 to 9). Other counters can readily be configured to count either up or down, because counting backwards to zero is fairly popular.

Before using a counter chip, I am going to design a counter to show how we apply what we already know about sequential logic design.

| Current Q2 Q1 Q0 | | | Next Q2 Q1 Q0 for | |
|---|---|---|---|---|
| | | | X=0 | X=1 |
| 0 | 0 | 0 | XXX | XXX |
| 0 | 0 | 1 | 001 | 010 |
| 0 | 1 | 0 | 010 | 011 |
| 0 | 1 | 1 | 011 | 100 |
| 1 | 0 | 0 | 100 | 101 |
| 1 | 0 | 1 | 101 | 110 |
| 1 | 1 | 0 | 110 | 001 |
| 1 | 1 | 1 | XXX | XXX |

FIGURE 7.1-1: State/output table using D.

## 7.1.1   Count 1 to 6

Design a counter that counts in binary from 1 to 6 and then repeats, but only when input $X = 1$. To say this another way, count 001–010–011–100–101–110–001–etc. while the input $X = 1$. Hold at the present count when $X = 0$.

We'll follow the design steps of Section 6.3:

1.  The word problem is pretty clear. In fact, counters generally don't lead to many questions.

2.  The state diagram isn't too useful because it is so obvious. There would be six circles, one for each of the six counts. The paths will go from one to the next for $X = 1$ and will loop to the same state for $X = 0$.

3.  The state/output table is where the design really starts. This is shown in Fig. 7.1-1. Notice that I have condensed the table a little by writing the next states in two columns, one for $X = 0$ and one for $X = 1$. If I wish to "read" the table into a Karnaugh map, I read by rows. There is no output column because there is no output specified.

4.  State minimization can't be done because the six states are specified.

5.  State variables are already assigned by the required counts.

6.  I *do* get to choose the type of flip-flop, though, since nobody has said what it must be. I am going to do the design twice, one with D and once with JK flip-flops. Recall that the excitations for D flip-flops are the same as the next-state values, so the table of Fig. 7.1-1 gives that information.

7.  The excitation table is done since I'm using D.

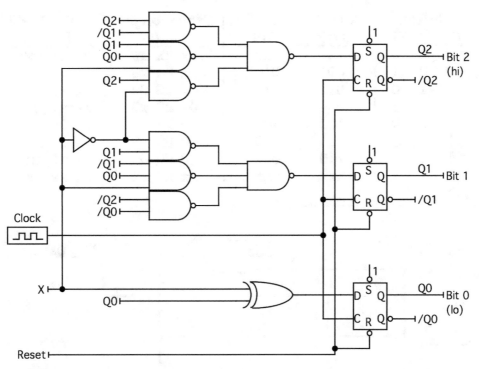

**FIGURE 7.1-2:** 1-to-6 counter using D flip-flops.

8.  I used Karnaugh maps to get the excitation functions:

$$D_2 = Q_2 \bullet \overline{Q_1} + Q_1 \bullet Q_0 \bullet X + Q_2 \bullet \overline{X},$$
$$D_1 = Q_1 \bullet \overline{X} + \overline{Q_1} \bullet Q_0 \bullet X + \overline{Q_2} \bullet \overline{Q_0},$$
$$D_0 = \overline{Q_0} \bullet X + Q_0 \bullet \overline{X} = Q_0 \oplus X.$$

9.  The finished circuit is shown in Fig. 7.1-2.
10. I tested it and the counter counts as required.

Repeating this using JK flip-flops requires me to expand my table to include six excitation functions, two for each of the three flip-flops. The expanded table is shown in Fig. 7.1-3. The excitation functions are

$$J_2 = Q_1 \bullet Q_0 \bullet X, K_2 = Q_1 \bullet X,$$
$$J_1 = Q_0 \bullet X, K_1 = Q_0 \bullet X + Q_2 \bullet X,$$
$$J_0 = X, K_0 = X.$$

| Current Q2 Q1 Q0 | Next Q2 Q1 Q0 for X=0 | X=1 | Excitation J2 K2  J1 K1  J0 K0 X=0 | X=1 |
|---|---|---|---|---|
| 0  0  0 | XXX | XXX | XX XX XX | XX XX XX |
| 0  0  1 | 001 | 010 | 0x 0x x0 | 0x 1x x1 |
| 0  1  0 | 010 | 011 | 0x x0 0x | 0x x0 1x |
| 0  1  1 | 011 | 100 | 0x x0 x0 | 1x x1 x1 |
| 1  0  0 | 100 | 101 | x0 0x 0x | x0 0x 1x |
| 1  0  1 | 101 | 110 | x0 0x x0 | x0 1x x1 |
| 1  1  0 | 110 | 001 | x0 x0 0x | x1 x1 1x |
| 1  1  1 | XXX | XXX | XX XX XX | XX XX XX |

FIGURE 7.1-3: State/output table using JK.

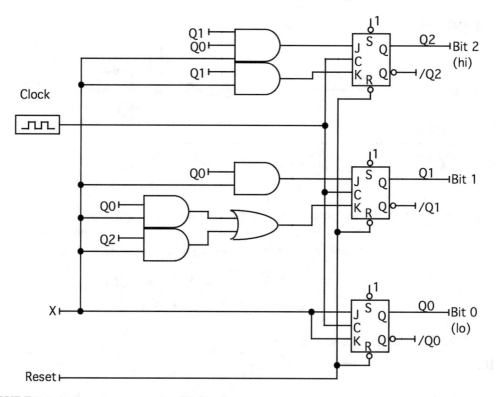

FIGURE 7.1-4: 1-to-6 counter using JK flip-flops.

Figure 7.1-4 shows the circuit, which has simpler gating than with the D flip-flops. But you can't generalize about D versus JK. They both have advantages.

My six-counter can be done with a single counter chip, the 74LS163, a four-bit counter shown in Fig. 7.1-5. The inputs allow counting, holding, and loading, as well as cascading chips for more bits. Here's what they do:

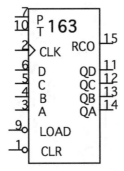

FIGURE 7.1-5: 74LS163 counter.

CLK is the clock. The counter counts on the rising edge of this clock.

CLR is the clear input, which forces the count to 0 on the next rising edge of the clock. (This
is called a *synchronous clear*.) This input is active low.

LOAD loads the counter with the value placed on the inputs D, C, B, and A, where D is the
high-order bit. This happens at the next rising edge. This input is active low.

P and T are count enables. Both must be asserted for the counter to count.

D, C, B, and A are the parallel-load inputs.

QD, QC, QB, and QA are the counter's outputs. QD is the high-order bit.

RCO is the *ripple count out* and becomes 1 when the counter is "full," which means $1111 = 15$.
This signal can be passed on to another '163 to make a counter with more bits. Keep in
mind that nothing takes effect except at the time of the rising edge of the clock. Since
RCO doesn't become 1 until after the current clock tick has started, it has no effect on
the next stage of a larger counter until the *next* clock tick.

Using this to make my 1-to-6 counter is fairly easy. I need a clock. I use CLR to provide
a Reset line. I use the input X to assert both enables, P and T.

The Load input is the interesting one. I want the counter to start at 001, which in four
bits is 0001. So I must arrange a signal from the output count that will load 0001 when the
count reaches $6 = 0110$. To do this I'll use an And gate to catch QC and QB when they both
become 1s. I don't need to look at any other bits because these two don't become 1s until the
count reaches 0110. This And gate really needs to be a Nand gate to provide the inversion
needed for the active-low Load input.

Finally, I must provide the value to load through the parallel-load inputs: $D = C = B =$
$0, A = 1$.

The Load input is synchronous, that is, it loads a new value on the next clock tick. The
counter therefore switches smoothly from 110 to 001. If Load were asynchronous (some devices

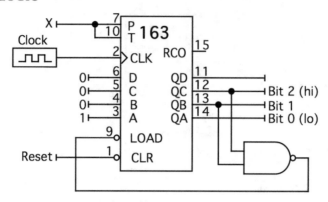

FIGURE 7.1-6: 1-to-6 counter using '163.

have such inputs), the counter would reset to 001 immediately upon reaching 110, not waiting for a clock tick.

The result is shown in Fig. 7.1-6, and simulation says it works correctly.

### 7.1.2    Count 30

Well, how about from 0 through 29? This will take two '163 chips cascaded. This time there's nothing needed to enable or stop the counter and there's nothing to load. So I attach a clock to both, pull Load up to disable parallel loading, and ignore the parallel inputs.

The interesting parts of the circuit, as shown in Fig. 7.1-7, involve "carrying" and catching the count at the end. I carry by using RCO of the low-order counter to enable the count of the high-order counter. This means the upper counter counts on the next clock tick after the lower one "becomes full." (The lower counter is enabled all the time.)

The end of the count is $29 = 11101$. I have used a five-input Nand gate to provide a low-going CLR signal to clear both counters on the next clock tick. That means the count after 11101 is 00000. Actually, this can be done with a four-input Nand, but I'll leave that to a problem for you.

Simulation says this counter works correctly.

### 7.1.3    "Gray" counting

A common binary code is a *Gray code* or a *Gray-like code*. These codes have the property that, when you are counting in them, only one bit changes at each transition. (These were discussed in Section 2.5.)

Design a counter that counts 6 in the Gray-like code 000-001-011-111-110-100-repeat. The counter is to count "up" through the sequence if an input $F = 1$; it is to count "down" if $F = 0$.

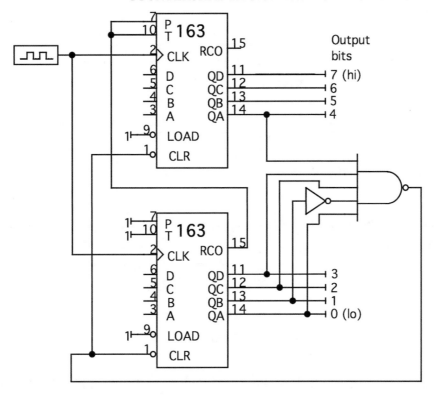

FIGURE 7.1-7: Count 30.

| Current | | | Next Q2 Q1 Q0 for | |
|---|---|---|---|---|
| Q2 | Q1 | Q0 | F=0 | F=1 |
| 0 | 0 | 0 | 100 | 001 |
| 0 | 0 | 1 | 000 | 011 |
| 0 | 1 | 0 | xxx | xxx |
| 0 | 1 | 1 | 001 | 111 |
| 1 | 0 | 0 | 110 | 000 |
| 1 | 0 | 1 | xxx | xxx |
| 1 | 1 | 0 | 111 | 100 |
| 1 | 1 | 1 | 011 | 110 |

FIGURE 7.1-8: Gray-like counter table.

The state table is shown in Fig. 7.1-8. I have written it in true binary order, even though that isn't the count order. If I don't follow the binary order, I will have a hard time "reading" the bits into Karnaugh maps.

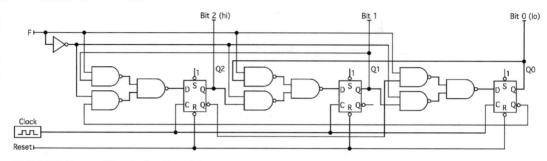

**FIGURE 7.1-9:** Circuit for Gray-like counter.

I have chosen to do this with D flip-flops, so the next-state values are also the D excitations. Using maps, I get

$$D_2 = Q_1 \bullet F + \overline{Q_0} \bullet \overline{F},$$

$$D_1 = Q_0 \bullet F + Q_2 \bullet \overline{F},$$

$$D_0 = \overline{Q_2} \bullet F + Q_1 \bullet \overline{F}.$$

The resulting circuit, which works in the simulator, is shown in Fig. 7.1-9.

I also tried this counter with JK and with T (trigger) flip-flops. Both are more complicated.

How about doing this with a counter like the '163? No, I am not even going to try! Standard counters are designed to count in fairly fixed sequences, usually binary. "Gray" counting isn't within the repertoire of these counters.

That's about enough on counters. They'll reappear in designs later, though. Now it's time to look at shift registers and what they can do.

## 7.2    SHIFT REGISTERS

Shift registers shift. All registers, shift or not, have a fixed length. Four bits, a byte, two bytes, and so on. So in a given situation, the register holds just a certain number of bits. Shifting these bits means moving them relative to their positions in the register.

If we think of a register as holding a binary number, then the bits can be moved either left or right. Moving them to the left, toward the high-order end of the register, is equivalent to multiplying by 2 on each shift. If you were doing this in decimal, this would be multiplying by 10. Moving bits to the right is equivalent to dividing by 2.

But the register is a fixed length, so if we shift either way, a bit must "fall off" one end and a new bit must "enter" the other end. That's where the shift register becomes useful.

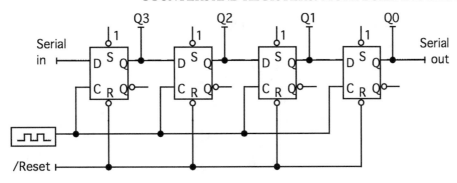

FIGURE 7.2-1: Simple four-bit serial-in, parallel/serial-out.

We have already seen the loading of a register with a group of bits. This is *parallel* loading. We also can read out all of the bits at the same time. A register like this would be designated as *parallel-in parallel-out.*

But we can also "slide" bits in from one end and "slide" them out the other. This serial bit stream is moving relative to the register. A register like this would be designated *serial-in serial-out.*

Our interest, though, is in the other two possible combinations: *serial-in parallel-out* and *parallel-in serial-out.* What for?

A good example is in a modem. Your computer works primarily in parallel when it is computing. But bits must be sent through the modem in serial fashion, one by one. So a byte is passed to the modem in parallel form, all eight bits at once. Then it is shifted out into the communications system one bit at a time. At the other end the process is reversed.

The register shown in Fig. 7.2-1 is a simple four-bit serial-in parallel-out register. Bits arrive at the left end. At each clock tick the bits in the D flip-flops are moved to the right and a new bit is received on the left. After four ticks of the clock four new bits from the serial bit stream occupy the register. These can be read out in parallel from the outputs Q3–Q0. Notice that the rightmost bit (which was the first to enter from the left) is also available on the serial output, so with three more clock ticks the remaining bits will appear there, one by one.

## 7.2.1 Universal Shift Register

The standard building blocks have a number of different shift registers available. I am going to consider just one, the 74LS194 universal shift register. I do this because we can do just about anything we want with this register without getting into the details of differences between it and others.

The four-bit register is shown in Fig. 7.2-2. It has some fairly obvious inputs such as the CLK and the input CLR to clear the register (active low). It has four parallel inputs and four

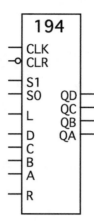

FIGURE 7.2-2: 74LS194 universal shift register.

| Function | Control | | Next | | | |
|----------|---------|-----|------|------|------|------|
| | S1 | S0 | QD | QC | QB | QA |
| Hold | 0 | 0 | QD | QC | QB | QA |
| Shift right | 0 | 1 | QC | QB | QA | R |
| Shift left | 1 | 0 | L | QD | QC | QB |
| Load | 1 | 1 | D | C | B | A |

FIGURE 7.2-3: Function table for 74LS194.

parallel outputs. L and R are the two serial inputs, one for each end. Control of the register is through S1 and S0.

The function table in Fig. 7.2-3 shows the four modes of operation. We can hold what is in the register, shift the contents to the right, shift them to the left, or load a new set of four bits through the parallel inputs.

This register is one example where the letter designations of the bits is backward. A *left* shift is from the D end to the A end. So making S1 S0 = 10 means to move the bit in QD to QC and so on, while taking in a new bit via the L input. (When you use this register, you'll probably mix up the directions at least once!)

What is all this good for? Let's do a counter example.

## 7.2.2   Ring Counters

My first shift register in Fig. 7.2-1 can be made into a ring counter. Aw, wow! So? What's a ring counter and why do I want one? A ring counter generally has one 1 bit that circulates through the counter register, moving over one position on each clock tick. When it gets to the end, it is reinserted at the beginning.

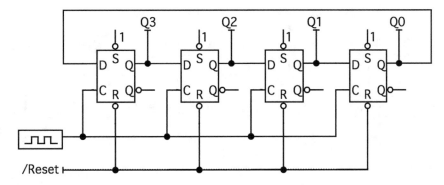

FIGURE 7.2-4:  Ring counter.

| Current | | | | Next | | | |
|---|---|---|---|---|---|---|---|
| Q3 | Q2 | Q1 | Q0 | Q3 | Q2 | Q1 | Q0 |
| 0 | 0 | 0 | 0 | 0 | 0 | 0 | 1 |
| 0 | 0 | 0 | 1 | 0 | 0 | 1 | 0 |
| 0 | 0 | 1 | 0 | 0 | 1 | 0 | 0 |
| 0 | 0 | 1 | 1 | 0 | 1 | 0 | 0 |
| 0 | 1 | 0 | 0 | 1 | 0 | 0 | 0 |
| 0 | 1 | 0 | 1 | 1 | 0 | 0 | 0 |
| 0 | 1 | 1 | 0 | 1 | 0 | 0 | 0 |
| 0 | 1 | 1 | 1 | 1 | 0 | 0 | 0 |
| 1 | 0 | 0 | 0 | 0 | 0 | 0 | 1 |
| 1 | 0 | 0 | 1 | 0 | 0 | 0 | 1 |
| 1 | 0 | 1 | 0 | 0 | 0 | 0 | 1 |
| 1 | 0 | 1 | 1 | 0 | 0 | 0 | 1 |
| 1 | 1 | 0 | 0 | 0 | 0 | 0 | 1 |
| 1 | 1 | 0 | 1 | 0 | 0 | 0 | 1 |
| 1 | 1 | 1 | 0 | 0 | 0 | 0 | 1 |
| 1 | 1 | 1 | 1 | 0 | 0 | 0 | 1 |

FIGURE 7.2-5:  Self-correcting ring counter.

Why? Many computer circuits need a series of timing pulses that activate various portions of the operation being performed. One pulse might let a byte into a register from a bus. The next pulse might start a multiplier, and so on.

I have converted Fig. 7.2-1 into a ring counter in Fig. 7.2-4. That didn't take much effort! But this circuit has a problem. What gets it started? It needs a single 1 somewhere. And what happens if it goofs? Suppose the 1 gets lost, or a second 1 appears somehow. How does it get corrected?

One way is to design a ring counter that is *self-correcting*. The transition table in Fig. 7.2-5 includes the four rows of proper operation (0001, 0010, 0100, and 1000). The bit in this

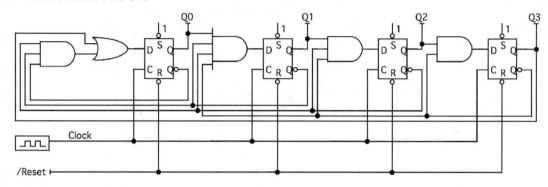

FIGURE 7.2-6: Self-correcting ring counter using D flip-flops.

case is shifting from the Q0 end toward the Q3 end. So 0001 goes to 0010 on the next clock tick, 0010 goes to 0100, and so on. But there are 12 other rows in the table, and all are possible if operation gets messed up somehow.

That's where the self-correction comes in. I have chosen a fairly simple algorithm to get the counter back on the track. For example, if the counter somehow gets into the 0011 state, the next value is 0100. The circuit is shown in Fig. 7.2-6.

A simple version of this ring counter (not self-correcting) is easily built with the '194 universal shift register. I merely get it started shifting with a single bit running through it. When the bit gets to QD, it is routed back to the input R. The circuit of Fig. 7.2-7 does the job. (I'm shifting to the right here.)

While this ring counter is not self-correcting, it does have a Load input that starts it with the correct value. Notice that Load $= 1$ loads because that makes S1 S0 $= 11$. Load $= 0$ allows the right shift to proceed at each clock tick.

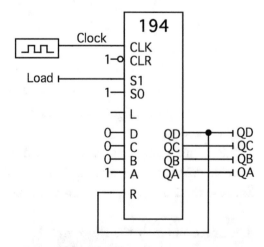

FIGURE 7.2-7: Ring counter, not correcting.

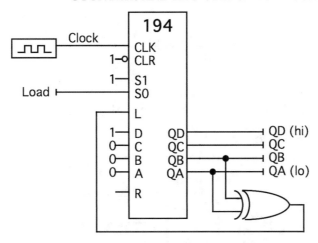

FIGURE 7.2-8: Pseudo-random counter.

### 7.2.3    Random Numbers

Would you believe that you can use a ring counter to create random numbers? You shouldn't believe this! After all, a digital system is entirely deterministic, so there can be no such thing as a truly random number. But pseudo-random counters are useful for creating bit patterns that aren't in any obvious sequential order.

These contraptions, also called *linear feedback shift-register* counters, are easy to design because the feedback necessary to make them work is well known. Any standard logic text will have a table of them. We aren't going to go that far, but I will mention two of these.

The generator is a shift register that moves a bit pattern at every clock tick. I will use the '194 in a shift-left configuration. For feedback, I use an exclusive-or gate to pick up the rightmost two bits (QB and QA) and feed them back to the serial input (L). The result is shown in Fig. 7.2-8.

What is the sequence? It is by definition 15 values long. But they occur in a "funny" order. So they appear random unless you are paying close attention.

Here are the feedback equations for some of the more common counter lengths. The bits are numbered from $Q_n$ to $Q_0$, where $Q_0$ is the "output end" of the shift register. In other words, the shift register is shifting *toward* $Q_0$.

| Length | Feedback |
|---|---|
| 4 | $Q_1 \oplus Q_0$ |
| 8 | $Q_4 \oplus Q_3 \oplus Q_2 \oplus Q_0$ |
| 16 | $Q_5 \oplus Q_4 \oplus Q_3 \oplus Q_0$ |
| 24 | $Q_7 \oplus Q_2 \oplus Q_1 \oplus Q_0$ |
| 32 | $Q_{22} \oplus Q_2 \oplus Q_1 \oplus Q_0$ |

| Current | | Next | |
|---|---|---|---|
| Q1 | Q0 | Q1 | Q0 |
| 0 | 0 | 0 | 1 |
| 0 | 1 | 1 | 0 |
| 1 | 0 | 1 | 1 |
| 1 | 1 | 0 | 0 |

FIGURE 7.2-9: Two-bit counter: D.

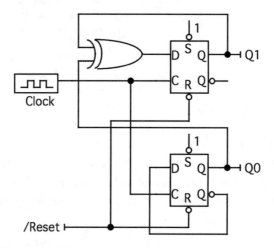

FIGURE 7.2-10: Two-bit counter: D.

### 7.2.4    Another Counter

Suppose, you say, I don't need a ring counter that complicated. And all this stuff about being self-correcting . . . . I'll just design a simple two-bit counter and then run the output through a decoder.

OK, fine, let's try it! Figure 7.2-9 shows the transition table. Using D flip-flops, the excitations are

$$D_1 = Q_1 \oplus Q_0, \, D_0 = \overline{Q_0}.$$

The circuit is shown in Fig. 7.2-10.

It's interesting here to try a JK implementation as well. The excitation table is shown in Fig. 7.2-11 and the excitations are

$$J_1 = K_1 = Q_0, J_0 = K_0 = 1.$$

The circuit is a simpler one, as Fig. 7.2-12 shows.

So now I will take the output and give it to a decoder as shown in Fig. 7.2-13. This should do the job. And it does. The only problem is that I can't find a standard decoder that

| Current | | Next | | Excitation | | | |
|---|---|---|---|---|---|---|---|
| Q1 | Q0 | Q1 | Q0 | J1 K1 | | J0 K0 | |
| 0 | 0 | 0 | 1 | 0 x | | 1 x | |
| 0 | 1 | 1 | 0 | 1 x | | x 1 | |
| 1 | 0 | 1 | 1 | x 0 | | 1 x | |
| 1 | 1 | 0 | 0 | x 1 | | x 1 | |

FIGURE 7.2-11: Two-bit counter: JK.

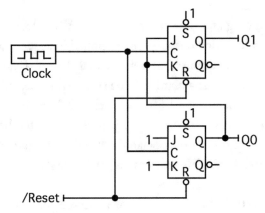

FIGURE 7.2-12: Two-bit counter: JK.

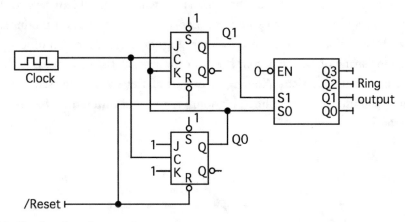

FIGURE 7.2-13: Another ring counter.

doesn't invert the outputs. Oh, well, maybe my ring counter will be OK with a zero running around instead of a 1.

This type of circuit can have interesting problems, though. It is possible, as the counter changes state, to get a brief but wrong output from one of the decoder outputs. That's because the counter does take some time to change and both flip-flops can never change exactly simultaneously. Hence, the counter can receive the wrong selection input for a brief moment.

You know about this kind of problem already in the form of a glitch. I had hoped to show a glitch when I simulated this circuit, but Murphy has prevented me from finding a combination and flip-flops and decoders that has such a glitch. The message here, I guess, is that glitches don't appear when you want them, only when you don't.

## 7.3    MEALY 'N MOORE

Are not a tag team in wrestling and they haven't played at Wimbledon. No, they are the names on two variants of the way in which outputs are developed in sequential circuits.

Let's go back to the first example of a sequential design, Example I of Section 6.4.1. Here's what I said I wanted. "A bit stream is to be searched for exactly two consecutive 1s. If such a pair is found, the output is to be 1 for one clock tick."

I started that design with a state diagram, Fig. 6.4-1, repeated here as Fig. 7.3-1. This is a Moore machine because the output is associated only with the state of the machine. It makes no difference what the inputs are. One we get to a particular state, the output is fixed by being in that state.

You can readily tell that the diagram is of a Moore machine—the output values are written in the circles that represent the states.

But you could argue that my design doesn't do what the problem says to do. You say that the problem requires "exactly two consecutive 1s." Yet the solution produces an output of 1 after two 1s, and if a third one comes along, it ignores that event. In other words, if the sequence is a long string of 1s, the output is 1 after the second, the fifth, the eighth, and so on.

If we accept that operation, allowing an output of 1 every time we see two 1s in a row, then the state diagram can be simplified by doing the output differently. Figure 7.3-2 shows this new approach.

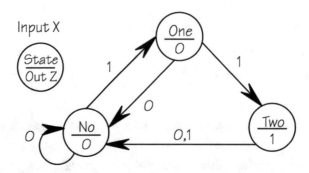

**FIGURE 7.3-1:**  State diagram of Fig. 6.4-1.

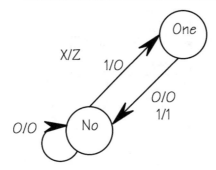

FIGURE 7.3-2: Mealy approach.

| Current Q | Input X | Next Q | Output Z |
|---|---|---|---|
| 0 | 0 | 0 | 0 |
| 0 | 1 | 1 | 0 |
| 1 | 0 | 0 | 0 |
| 1 | 1 | 0 | 1 |

FIGURE 7.3-3: Table for Mealy.

Notice the change in notation:

- There is no output associated to any state. In other words, the outputs aren't shown in the state circles.
- The labels on the transition paths now include two elements separated by a slash.
- The information before the slash is the input that causes that transition.
- The information after the slash is the output associated to that transition.

This is a *Mealy* machine, one in which the outputs are associated to the transitions. The output functions will involve not just the state variables (Qs) but the inputs as well.

So is this an improvement? It depends on what happens. In this case, the number of states has been reduced by one. More importantly, that number has gone from three, which requires two flip-flops, to two, which requires only one. That's probably better in some sense.

If I assign 0 and 1 to the No and One states, I get a transition/output table as shown in Fig. 7.3-3. The excitation and output functions are

$$D = \overline{Q} \bullet X,$$

$$Z = Q \bullet X.$$

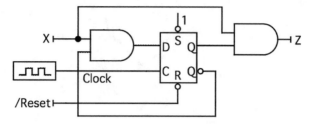

FIGURE 7.3-4:  Mealy circuit.

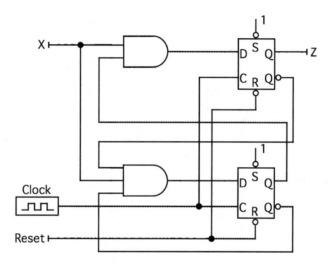

FIGURE 7.3-5:  Original Moore circuit.

The circuit is shown in Fig. 7.3-4. Notice the gating on the output includes the input X. Compare this with the original Chapter 6 solution, repeated in Fig. 7.3-5. That's an improvement, I'd say!

Of course, the thing still produces outputs for strings of input 1s. But I'll leave to you some of the other interpretations of what is wanted.

OK, so now you have a choice of Mealy or Moore when you design something. How do you know which is going to yield a simpler circuit by some measure? You don't.

Conceptually, Moore is a little easier to think about. You have states, and these states relate directly to desired outputs. But that doesn't say that we shouldn't try Mealy if we have any clue that it might improve the result.

I must admit that I have always had trouble remembering which M is which. My mnemonic is that since "Moore" has a double "o," this stands for "output only." Yeah, bad, but it works for me.

Enough of this stuff. Let's design something.

## 7.4    PARALLEL-SERIAL CONVERTER

Data are arriving on eight parallel lines labeled D7-D0. A strobe ST announces the availability of the next byte by a high-going pulse.

When ST occurs, accept the byte into a register.

On the first clock tick after ST, start a ten-bit shift to produce a serial output. The order of the output is to be

$$1 \; D0 \; D1 \; D2 \; D3 \; D4 \; D5 \; D6 \; D7 \; 0$$

The leading bit is a fixed 1 followed by the eight data bits. It ends with a fixed 0. Hence, the eight-bit parallel input becomes a ten-bit serial output.

There are lots of ways to approach this problem. One is to create a large state diagram and then go through the formal process of sequential logic design. But that will yield a very large table.

We will do better if we will use some of the building blocks that we already know about. Here are the steps I am going to follow:

1.  Design a 12-bit shift register using three '194 chips.
2.  Arrange this register to have a fixed 1 at the "bottom" end.
3.  When the strobe ST comes along, load the byte with D0 next to the 1 at the bottom end.
4.  Put a fixed 0 at the other end, just "above" D7.
5.  Design a counter to count ten events. Start this counter with the strobe ST. Also start the shifting in the same way.
6.  When the count is full, stop the shifting and reset the counter to 0.

My design begins with the shift register. Figure 7.4-1 shows three '194s cascaded to form a 12-bit register (although I'll use only ten of those). I am planning to shift from top to bottom, which is "left" in '194 parlance.

The inputs to the register are shown. Notice the fixed 1 and the fixed 0. The output is at the "bottom." The upper registers feed their outputs to the L inputs of the registers below.

The counter is shown in Fig. 7.4-2. Because the ST strobe is short, lasting only for one clock tick, I have used a JK flip-flop to remember that it has occurred. ST drives the J input of the flip-flop, creating the signal SH that will be used to enable the shift register.

The '163 counter is cleared directly by ST. Counting is then enabled by SH, which comes one clock tick later. When the counter reaches ten (1010), the And gate provides the Stop signal that resets the JK flip-flop and halts the whole process.

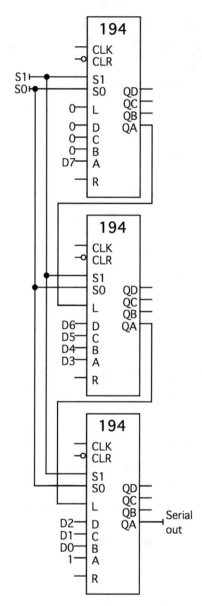

FIGURE 7.4-1: 12-bit shift register.

Now I need to develop logic to provide S1 and S0 for the shift register. The table in Fig. 7.4-3 shows this. I want to load on ST, so S1 S0 = 11. Then I want to shift on SH, so S1 S0 = 10. This requires S1 = SH + ST, S0 = ST.

Figure 7.4-4 shows the complete circuit. But does the circuit work? Of course! How could you doubt my work? I tested it and it *looks* good.

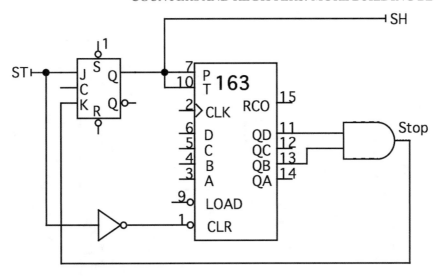

FIGURE 7.4-2:  Start/stop and 10-counter.

| SH | ST | S1 | S0 |
|----|----|----|----|
| 0  | 0  | 0  | 0  |
| 0  | 1  | 1  | 1  |
| 1  | 0  | 1  | 0  |
| 1  | 1  | x  | x  |

FIGURE 7.4-3:  S1 and S0.

But .... A careful test shows that the shifting seems to go one too many times. This can be verified by changing the fixed bit just above the fixed 0 to a 1.

Hmmm, the output looks OK but it isn't exactly right. Where's the problem? Did you catch it already? How come I am shifting eleven times instead of ten?

This is a common design error in sequential logic (and in computer program loops too). The counter is running one too far. I stopped the counter on 10, which means that it counts *eleven* times, from 0 through 10. (An arrow in Fig. 7.4-4 marks the location of the error.)

So I change the Stop signal gating to catch 9 = 1001 instead and the system works correctly. This is done by moving the wire at the arrow shown in Fig. 7.4-4 down one pin to QA.

This has been an example of some formal design (Fig. 7.4-3 for S1 and S0, for example) and some ad hoc design (the JK flip-flop and the counter, for example). This is the kind of approach that we generally take in designing larger systems.

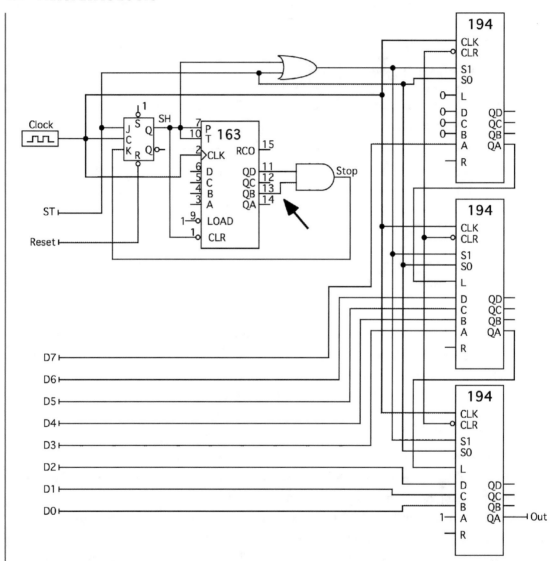

**FIGURE 7.4-4:** Parallel/serial converter.

## 7.5   SUMMARY

The larger building blocks are a big help in many sequential designs. Standard counters and shift registers simplify our work. Sometimes it takes being quite clever to adapt a design to a standard block, however.

Counters generally can be designed by going directly to the state table. The state diagram is too simple and the binary states have already been specified.

Shift registers can be used as counters that count in unusual patterns. We've seen both ring and pseudo-random counters using them.

Larger system can be broken up into small segments that we can handle more easily. The ten-bit shifter design shows this. Large systems can be very hard to design without breaking them up.

In the next chapter, I will do a modular design of a large system, the controller for a microwave oven.

CHAPTER 8

# Design a Microwave: Well, the Controller

OK, let's do it. We've been talking about logic circuits and design and such stuff for seven chapters. It's about time we designed something. How about designing a microwave oven? Uh, well, no, not really the microwave oven. After all, the heart of the microwave is the magnetron, and that was invented in 1940 by Hans Hollmann and played a major part in the Allied successes in World War II.

We'll design the controller, which is the "logic part" of the microwave. Can you do this now that you've had the beginnings of logic design? Possibly not, because we haven't looked at the design of a large system yet. That's what this chapter is about. I'll do it by designing the microwave oven controller.

What's the problem with this controller that is different from what we have been doing? Size. This problem is larger than anything we have done. The tools we have, such as Karnaugh maps, don't handle larger circuits well. So we need something better.

On top of that, large problems can't really be solved "all at once." Instead, we need to take the large problem and break it down into smaller parts. We keep doing this until the parts are small enough for us to work with. This means something different to different people. I should, for example, be able to handle a somewhat larger "part" than you can at the moment.

I need to warn you about how an example like this is developed for a text. Recall being in a math class and watching the instructor develop a proof. The proof was developed in a very orderly manner, often elegant. But did the first mathematician who developed that proof do it in such a perfect manner? Very unlikely! The path to the proof probably involved many twists and turns, blind alleys, wrong theorems, and so on. What you saw so beautifully presented in class was the result of lots of refinements.

While I will try to tell you what's been happening in my mind over the last couple of weeks as I developed the controller, I cannot completely succeed in doing that. Each piece of my presentation is a distillation of everything that I went through to get a working controller. I can tell you some of the twists and turns and blind alleys and wrong choices, but I can't tell everything.

OK, enough talking! What's going to happen? I'll start by defining the problem and developing the specifications. Then I'll give you a clue as to how I finally figured out how to start breaking the problem into pieces I could handle. After that, I'll design the pieces, test them, put them together, and test the whole thing.

## 8.1   SPECIFICATIONS

Design problems are often incompletely formed. The person asking for the design doesn't know quite everything about the system, words can be interpreted in more than one way, some difficulties are not foreseen, and so on. So we need *specifications*.

The specifications tie down all the details so that the designer knows what is to be designed. More important, though, is the use of the specifications to test the final design. Either the design meets those specifications or it doesn't. If it does, the designer gets paid; if it doesn't, it's back to the drawi..., well, today it's back to the computer.

So let's get some specifications. I will admit here that I did tailor the specs a little to keep the project manageable for this text. But what we'll get is a pretty complete controller for a low-to-mid-priced microwave oven.

### 8.1.1   Basic Functions

This microwave oven is to be able to cook something for a period of time. This time is set by the cook and the machine is started. At the end of the time specified, the oven stops cooking. That's pretty basic. But perhaps we could add a few features? Here are the basics:

- Three power settings, high, medium, and low.
- Entry of cook time in minutes and seconds from a ten-digit keypad.
- Safety switch on door that prevents cooking with the door open. Opening the door during cooking stops the cooking and holds but does not reset the cooking time.
- Stop button that allows the cook to stop the cooking. Pushing the button a second time aborts the run completely.
- Time-of-day clock that is displayed when the oven is not being used for cooking. Clock is set using the ten-digit keypad. Time display is in hours and minutes using 12-hour time.

Some things that often are not said still must be said. For example, what happens when you are setting the time-of-day clock and press the Start button? So that adds another specification:

- Random button pushings don't cause random or inappropriate events.

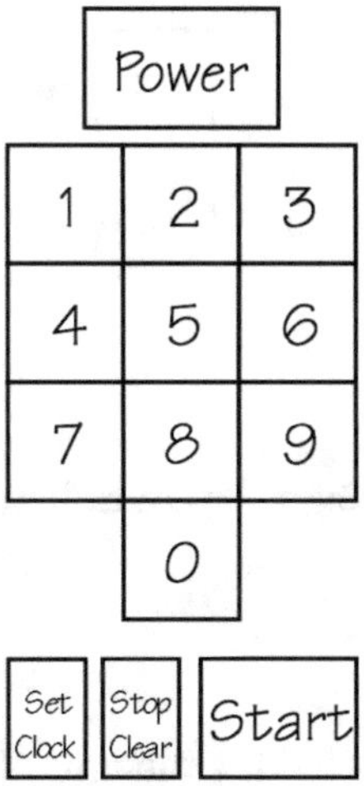

FIGURE 8.1-1: Keyboard.

Nothing's been said about where "time" comes from. After all, there must be a basic clock that ticks away at some standard speed. There are generally two possibilities. One is a quartz–crystal oscillator such as you might have in your wristwatch. But this oven is useful only when connected to commercial power, so the second time source seems like a better choice:

- "Time" comes from pulses derived from the 60-Hz power line (which over a day or so keeps nearly perfect time). This pulse frequency will be 60 pulses per second.

There! That pretty well tells what our controller is to do. Now how about the user interface, since that will determine a number of things for us.

## 8.1.2   User Interface

The user needs to be able to choose the power setting, set the cook time, start and stop the cooking, clear the cook time, and set the time-of-day clock. Figure 8.1-1 shows a possible arrangement.

The display needs to be simple in keeping with the rather simple operation. All we need is a way to display time-of-day, cook time, and power setting. The specifications don't mention how much time-of-day is to be displayed, so I will choose to display hours and minutes. This means that the cook time will be displayed in minutes and seconds. Figure 8.1-2 shows the display.

Notice that I have tied down one thing that the specs don't mention. The maximum cook time is 59:59, so if you want to cook something 60 minutes, you can't ... quite. Is this

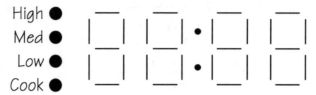

FIGURE 8.1-2:  Display.

OK with the person who told us to design the controller? This would be a good thing to check before going too far.

### 8.1.3   Some Additions

As I thought through what I wanted to do, I thought of four more specs that I didn't choose to implement:

- Three beeps when cooking is finished.
- Beep if a keyboard button is pressed before the power level is set.
- Beep if the user attempts to set a cooking time greater than 19:59.
- Beep if a time-of-day greater than 12:59 is entered.

Why did I develop these ideas and then delete them? Mostly to keep the problem from getting too big. Hmmm, but these might make interesting problems for you to do.

## 8.2    SYSTEM DESIGN

The first step in doing this design is to get the big picture of how things flow, how they fit together, what influences what, and so on. So that's what I did. I started thinking through various button-pushing sequences and developed a State Diagram. Figure 8.2-1 shows a scan of my very sloppy diagram. It has numerous erasures and some other problems as well. (Don't try to read it in detail!)

But I quickly realized that I was not getting to the heart of the system. My diagram didn't represent anything that I could reduce to working logic. Worse than that, it had many states (close to a dozen) that were beginning to seem too complicated to me. I also began to find that I had paths missing and some paths that I couldn't clearly define. I needed a new approach.

### 8.2.1   Finite State Machine

That new approach is through what is called a Finite State Machine, or FSM for short. Instead of trying to think about the controller as a whole and find various states and their transitions, I needed to think about the states the controller could find itself in.

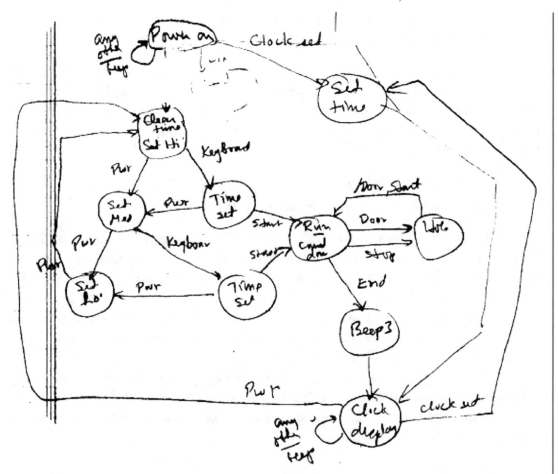

**FIGURE 8.2-1:**  Poor first try for controller.

When my brain began to function properly, I realized that this microwave-oven controller had five different states. (Recall that a *state* is one set of conditions on all the memory elements, whatever they might be. Remember standing up and sitting down?) Once I know these states, I can think about what gets the FSM from one to another.

Here are the five:

- Initial: The state the machine starts in when the power comes on.
- Set clock: Before anything else can happen, the time-of-day clock must be set, which is done in this state. (I'll call this simply the *clock* from now on.)
- Ready: The machine is displaying the clock and ready to receive instructions.
- Run: The cook time is set and the cooking takes place in this state.

- Hold: The cooking is stopped by the Stop button or by opening the door; the time on the cook timer is held. (I'll call this just the *timer* from now on.)

I started working with these states to determine what moves my Finite State Machine from one to another by drawing a State Diagram and sketching the paths. But it wasn't long before I found that I needed one more state. The *Run* state in the list above is too complex because it contains two major events: setting the timer and cooking by counting down the timer. So I replaced this state with two:

- Set timer: Set the cooking time.
- Cook: Run the microwave and count the timer down.

That gives me six states, which is a number that I can manage as I continue the design. Note that I am not saying that my oven has only six states! In fact, it has a humongous number because each tick of the clock is a different state (1:34 is not the same state as 1:35, for example).

Within each of these states there will be states. But I am going to design the Finite State Machine to provide these six states. I'll use signals from these states to determine what actions certain parts are to have. For example, during Set clock and Ready the four-digit display shows the clock, but during Set timer, Cook, and Hold it shows the timer.

How do we get from one state to another? The next step in the design is to develop the *transitions* that govern these paths. If I have done a good job of defining the states, the transitions won't be complex. The transitions are caused by signals from two sources, the user interface and conditions within the oven.

### 8.2.2   Signals

My user interface has four signals that are going to be part of the transitions:

- Clock: This button allows the user to start setting the clock. Pushing it again means the user has set the time. I'll call this signal Clk_Set.

- Power: This button will step through the three power settings, high, medium, and low. I'll name this Pwr.

- Start: This button starts the cooking after the timer is set. I'll name it Start.

- Stop-Clear: The first push of this button interrupts the cooking; the second terminates it. I'll name it Stp_Clr.

Now here's a clue that you might overlook at the beginning. These button pushings must be *synchronized* to the system clock. Keep in mind that we are designing a synchronous sequential circuit, which means there is a system clock. This clock is *not* the time-of-day clock or the cook timer. It is the internal clock that drives all the flip-flops. In this controller it will be the 60-Hz pulses.

But the button pushings won't be timed by this clock, since the user can press a button any time. In addition, the buttons probably have to be *debounced* so that their mechanical bouncing isn't read as multiple pushes.

Moreover, I can make my transitions simpler if I can guarantee that the button signals don't last for as long as the user holds the button down. So I will run the button signal through a *one-shot* that produces a single pulse shaped just like a single clock tick.

Two other signals are needed, signals that will come from within the machine:

- Door: This switch tells that the door has been opened. I'll name it Door.
- Done: This signal indicates that the timer has run to zero and cooking is finished. I'll call this Run_Dun.

There! Now I have the signals so I can work on the transitions.

## 8.2.3   Transitions

The State Diagram is shown in Fig. 8.2-2. It shows how the FSM gets from one state to the next. As I drew the diagram, I put in only the circles (states) and their names. The binary state values came much later as I'll explain. The arrows follow from thinking through the operation of the microwave:

*Initial* is the starting state and happens when power is applied. This state is generally assigned 000 because many flip-flops either have a separate clear input or come on in the zero state.

The only allowed activity from *Initial* is setting the clock, so the transition is the signal Clk_Set. Once in *Set clock* the only exit is via a second Clk_Set to indicate that setting is finished.

*Ready* is the normal idle state. It has two exits. One is Clk_Set so the user can redo the clock. The other is the beginning of the cooking setup. I have used Pwr as the transition to the next state because the user must select a power level before setting the cook time. Note that this does not prevent further pushes of the Power button; it says only that this first push causes the transition.

In the *Set timer* state, the user is setting the cooking time. One exit is the Stop button (Stp_Clr) so the user can abort the operation. The other is to the *Cook* state via the Start button. But in addition, the door must be closed, so the transition is Start • Door' (Start and Not Door).

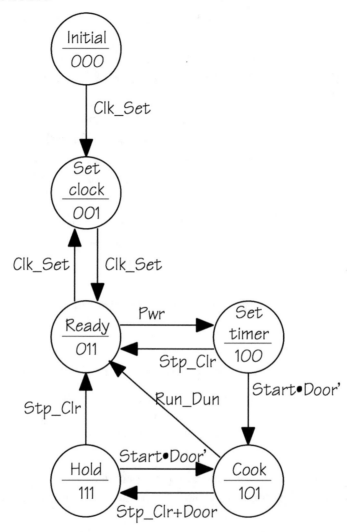

FIGURE 8.2-2: State Diagram.

The *Cook* state has two exits. One occurs when cooking is finished and the timer runs out: Run_Dun. The other is either a push of the Stop button (Stp_Clr) or the opening of the door. So the transition is Stp_Clr + Door.

From the *Hold* state there are also two exits. If the user pushes the Stop button again, the run is aborted. Pushing Start with the door closed returns the oven to the *Cook* state.

I hope you notice that this diagram for the Finite State Machine is simple and orderly! It's certainly a far cry from the first diagram (Fig. 8.2-1) that I fumbled with.

| Current state | Current Q2 Q1 Q0 | Transition Functions | Next state | Next Q2 Q1 Q0 |
|---|---|---|---|---|
| Initial | 0 0 0 | Clk_Set | Set Clock | 0 0 1 |
| | 0 0 0 | Clk_Set' | Initial | 0 0 0 |
| Set Clock | 0 0 1 | Clk_Set | Ready | 0 1 1 |
| | 0 0 1 | Clk_Set' | Set Clock | 0 0 1 |
| Ready | 0 1 1 | Pwr | Set Timer | 1 0 0 |
| | 0 1 1 | Clk_Set | Set Clock | 0 0 1 |
| | 0 1 1 | Pwr' • Clk_Set' | Ready | 0 1 1 |
| Set Timer | 1 0 0 | Start • Door' | Cook | 1 0 1 |
| | 1 0 0 | Stp_Clr | Ready | 0 1 1 |
| | 1 0 0 | (Start • Door')' • Stp_Clr' | Set Timer | 1 0 0 |
| Cook | 1 0 1 | Run_Dun | Ready | 0 1 1 |
| | 1 0 1 | Stp_Clr + Door | Hold | 1 1 1 |
| | 1 0 1 | Run_Dun' • (Stp_Clr + Door)' | Cook | 1 0 1 |
| Hold | 1 1 1 | Stp_Clr | Ready | 0 1 1 |
| | 1 1 1 | Start • Door' | Cook | 1 0 1 |
| | 1 1 1 | Stp_Clr' • (Start • Door')' | Hold | 1 1 1 |

FIGURE 8.2-3: State/Transition Table.

The state variable assignments are sort of ad hoc, although I did think about some of them. 000 is obvious, and I simply went to 001 from there. From 001, 011 is a single bit change, which sometimes produces simpler logic. I chose 100 next just because I decided to use the leftmost bit for all three cycles related to cooking. Then the remaining two cycles are single-bit changes around the loop.

Would other assignments have yielded simpler logic? There's only one way to find out, which is to try others. And that is a good source of homework!

The State/Transition Table of Fig. 8.2-3 merely recites the State Diagram in tabular form. I include the names of the states so I don't get confused.

One thing the diagram doesn't show, though, that must be in the table is the conditions for remaining in a given state. Note, for example, that Clk_Set moves the machine from 000 to 001. If that's the only way to get from 000 to 001 (which it is), then Clk_Set' is the way to stay in 000. You'll see this throughout the table. First I write the transitions and then I use their complement as the "stay here" condition.

## 8.2.4   Excitations

Now comes a tricky part. I can write the excitation functions directly from the table. But they have lots of variables, namely, three Qs and six signals or conditions. That's nine variables! A 512-line truth table? Not likely! But I can still write the equations.

For what type of flip-flop? I have chosen D. Perhaps the excitations would be simpler using JK, but I decided they wouldn't. Arbitrary, perhaps? Perhaps!

The equation for $D_2$ is given below. I have used digits like "3" for the state value rather than Qs to save writing. Rather long, eh?

$$D_2 = (3) \bullet Pwr + (4) \bullet Start \bullet \overline{Door} + (4) \bullet (\overline{Start} + Door) \bullet \overline{Stp\_Clr}$$
$$+ (5) \bullet (Stp\_Clr + Door) + (5) \bullet \overline{Run\_Dun} \bullet \overline{Stp\_Clr} \bullet \overline{Door}$$
$$+ (7) \bullet Start \bullet \overline{Door} + (7) \bullet \overline{Stp\_Clr} \bullet (\overline{Start} + Door)$$

How do you simplify such an equation without writing a huge truth table? One simplification is to ignore simplifying on the basis of the Qs. I plan to use a 8-to-1 multiplexer to provide the flip-flop input anyhow, so simplifying the Qs doesn't help anything. (Using a multiplexer generally reduces the number of individual gates.)

I can probably simplify this function by combining terms related to the same state, though. That means that the first term can't be reduced because there is no other "3" terms to combine with. But there are two "4" terms and I can use a simple map involving only Start, Door, and Stp_Clr to reduce this to one "4" term with some gating. Similarly, "5" and "7" terms can be reduced through simple maps that deal with only the signals that appear with those states. The resultant simplified $D_2$ is

$$D_2 = (3) \bullet Pwr + (4) \bullet (\overline{Stp\_Clr} + Start \bullet \overline{Door})$$
$$+ (5) \bullet (Stp\_Clr + Door + \overline{Run\_Dun})$$
$$+ (7) \bullet (\overline{Stp\_Clr} + Start \bullet \overline{Door})$$

Without further comment I'll show the remaining excitation functions and their simplifications:

$$D_1 = (1) \bullet Clk\_Set + (3) \bullet (\overline{Pwr \bullet \overline{Clk\_Set}}) + (4) \bullet Stp\_Clr$$
$$+ (5) \bullet Run\_Dun + (5) \bullet (Stp\_Clr + Door) + (7) \bullet Stp\_Clr$$
$$+ (7) \bullet \overline{Stp\_Clr} \bullet (\overline{Start} + Door)$$
$$D_1 = (1) \bullet Clk\_Set + (3) \bullet \overline{Pwr} \bullet \overline{Clk\_Set} + (4) \bullet Stp\_Clr$$
$$+ (5) \bullet (Run\_Dun + Stp\_Clr + Door)$$
$$+ (7) \bullet (Stp\_Clr + \overline{Start} + Door)$$

$$D_0 = (0) \bullet Clk\_Set + (1) \bullet Clk\_Set + (1) \bullet \overline{Clk}\_Set + (3) \bullet Clk\_Set$$
$$+ (3) \bullet \overline{Pwr} \bullet \overline{Clk}\_Set + (4) \bullet Start \bullet \overline{Door} + (4) \bullet Stp + Clr$$
$$+ (5) \bullet Run\_Dun + (5) \bullet (Stp\_Clr + Door)$$
$$+ (5) \bullet \overline{Run}\_Dun \bullet \overline{Stp}\_Clr \bullet \overline{Door} + (7) \bullet Stp\_Clr$$
$$+ (7) \bullet Start \bullet \overline{Door} + (7) \bullet \overline{Stp}\_Clr \bullet (\overline{Start} + Door)$$
$$D_0 = (0) \bullet Clk\_Set + (1) + (3) \bullet (Clk\_Set + \overline{Pwr})$$
$$+ (4) \bullet (Stp\_Clr + Start \bullet \overline{Door}) + (5) + (7)$$

### 8.2.5   FSM Circuit

An 8-to-1 multiplexer can provide a simple way to drive each of the D flip-flops. The output of the mux is the input of the flip-flop. The selection inputs of the mux are $Q_2$ $Q_1$ $Q_0$ and the mux is permanently enabled.

Figure 8.2-4 shows the circuit for the Finite State Machine. There are three muxes. Notice that the gating on the mux inputs is exactly what is in the simplified excitation functions. I did not draw all the wires. Instead, I used signal names to link terminals.

The state outputs are needed as individual signals. In other words, I need a separate signal that is asserted when the FSM is in a particular state. A 3-to-8 decoder is a simple way to do this.

I included in the FSM the one-shots needed to synchronize the signals from the buttons. These are on the left in Fig. 8.2-4. They insure that a button push is cleaned up to become a single pulse synchronized with the system clock.

One of the purposes of modularizing the circuit is to make the modules small and simple so that they can be tested 100%. This makes development less time consuming because testing time is reduced. Then when the final circuit is put together, you don't have to debug hundreds of little parts.

The circuit of Fig. 8.2-4 has *ports* added to it. Those on the left are inputs; those on the right are outputs. I then combined this whole circuit into one modulethat has my signal labels around the edges and all the logic hidden inside (Fig. 8.2-5).

## 8.3   COUNTERS AND TIMERS

My controller involves counting to provide time. Developing these counters takes some thinking because there are lots of ways to do the job. But the first step is to figure out what must be counted. When I considered both the clock (time-of-day) and the timer (cooking), I realized that I had to count three different things:

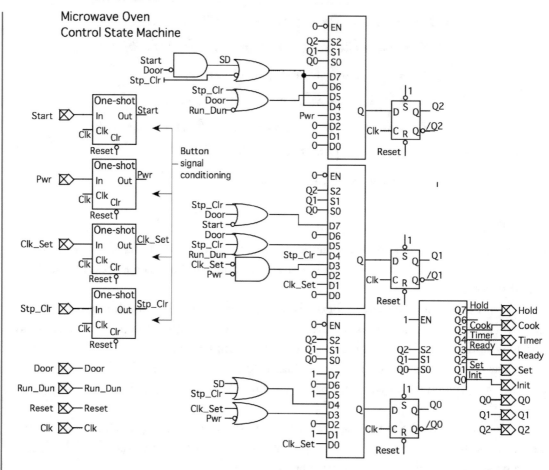

FIGURE 8.2-4: Control Signals State Machine.

Seconds from 0 to 59,
Minutes from 0 to 59,
Hours from 1 to 12.

Yet at any given time only two of these three have to be active. Or at least that is what I thought! Wrong, Bill! You can't just quit the clock when the timer is running. You need both a clock (0-59, 1-12) and a timer (0-59, 0-59). While they both will use the same display, only one is displayed in any given state.

Another thing that I quickly realized is that the counters need to be in BCD (binary-coded decimal) so that I don't have to change from binary to decimal to display the times. Hence, I'll need single-digit counters that count 0-5, 0-9, and 1-12.

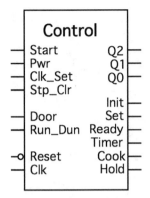

FIGURE 8.2-5: State Machine as part.

I played for a long time with standard parts, up counters, up–down counters, and flip-flops. I couldn't find a simple counter that could do 0-5, 0-9, and 0-59. In addition, the clock must count up and the timer must count down.

Another factor that I considered is how to set the clock and the timer. One way is through the usual parallel load that we have seen on standard counters. The counter has a Load input that accepts a new setting from a group of input lines. So I initially included a parallel load in my counter design

However, I quickly realized that this was not the manner in which the counters would be set in practice. Instead, the user would press, say, the Set-clock button, then enter the digits on the keypad. As these digits are entered, the display shifts them across from the right to the left. So my counters really needed shift inputs.

What are shift inputs? Instead of parallel load, I "parallel shift" the BCD digits. Asserting the Shift input of the counter transfers BCD digits to the next positions to the left and accepts a new digit on the right.

## 8.3.1  Up Counters

I decided to work on the clock first, which counts up. I need three counters, up 0-5, up 0-9, and up 1-12, all with the ability to shift BCD digits to the left. They also need carry-in and carry-out terminals so they can be linked to make larger counters. All need to be four-bit BCD counters even though counting 0-5 requires only three bits. This keeps the display simpler and also allows digits larger than 5 to shift through the BCD position.

I chose to use JK flip-flops for these counters, mainly because counters tend to be a little simpler with JK. If I want the bit to be 1, I assert J; if I want 0, I assert K.

The up counter for 0-5 is shown in Fig. 8.3-1. Let's look more closely at the logic for the flip-flop Q2:

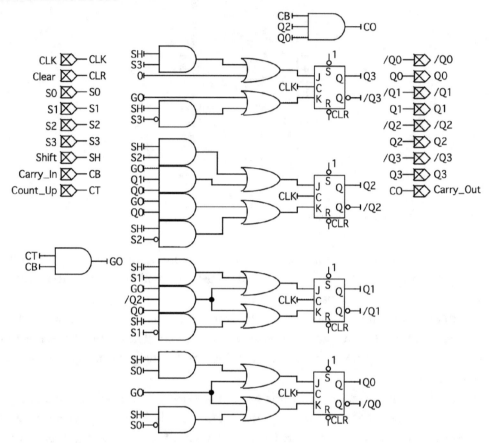

**FIGURE 8.3-1:** Up counter: 0–5.

- Look first at the middle two And gates that derive the Q2 flip-flop. Both are active only when the GO signal is asserted. The GO signal is provided by the And gate on the left edge using the Count_Up and Carry_In signals together.

- Notice that J2 = GO•Q1•Q0 and that K2 = GO•Q0. I got these by doing a standard counter design, writing a table of Current $Q_2Q_1 Q_0$ and Next $Q_2 Q_1 Q_0$, then developing J and K, and finally adding these with GO.

- Now look at the inputs S2. This is the input of the BCD digit being shifted from the right. If S2 = 1, then J = 1; if S2 = 0, then K = 1. The control signal SH activates these inputs.

After testing this circuit, I converted it into a module named Up-5 Counter for use later.

Design of the up counter for 0-9 (Fig. 8.3-2) and the up counter for 1-12 (Fig. 8.3-3) are done in much the same way. I made each of these counters into a module after testing each one completely.

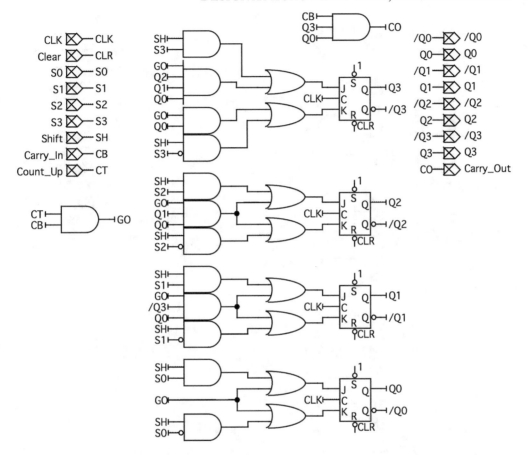

FIGURE 8.3-2:  Up counter: 0–9.

But do they work as a clock? I put the three counters together as shown in Fig. 8.3-4 to try them. After completely testing this combination, I combined these counters into a larger module, the 12-Hour Clock.

In Fig. 8.3-4, notice the linking of the shifting digits. The outputs of the up-9 counter become the shift inputs of the up-5 counter, and so on. I also converted the shift signal into a single pulse so that I would have just one shift no matter how long the shift signal is held.

## 8.3.2   Down Counters

Whew, and ugh! Gotta do this all over again for the timer, which counts down. This counter needs a Down-5 Counter and a Down-9 Counter twice. It also needs the same shift arrangement. Figures 8.3-5 and 8.3-6 show the basic counters, and Fig. 8.3-7 shows the resultant timer, which then became, after testing, the Timer module.

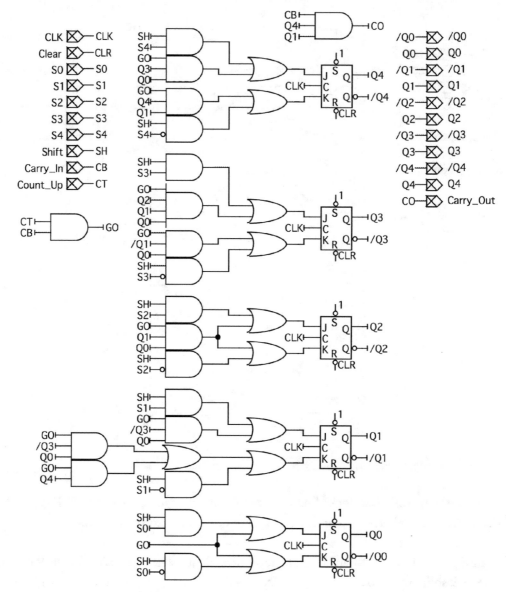

FIGURE 8.3-3: Up counter: 1–12.

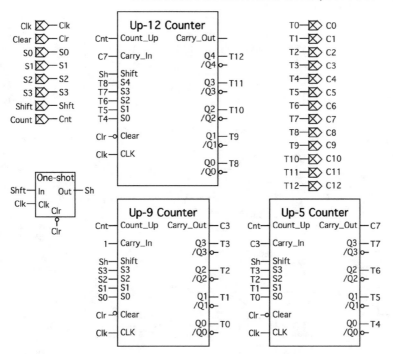

**FIGURE 8.3-4:** 12-hour clock.

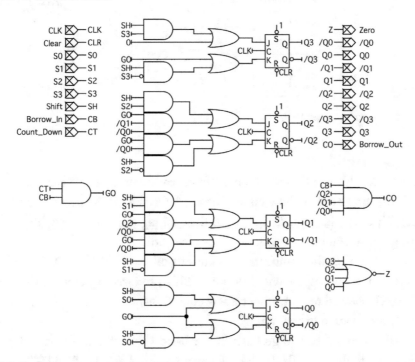

**FIGURE 8.3-5:** Down counter: 5–0.

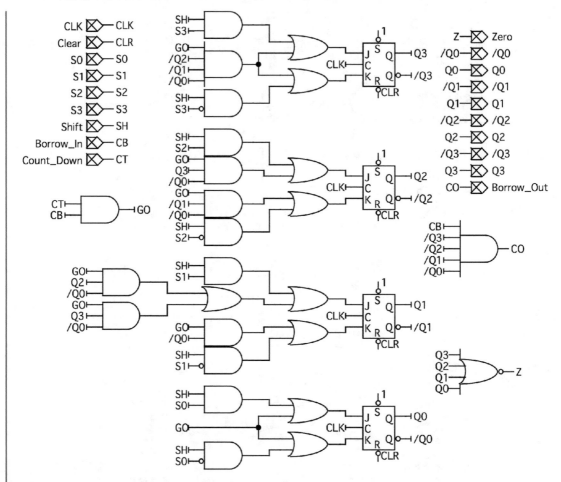

**FIGURE 8.3-6:** Down counter: 9–0.

### 8.3.3   Do They Work?

By now you are probably wondering whether things really worked this well? Yes, mostly, but a problem with this timer didn't show up until much later. The FSM needs to know when the timer reaches 0 because that signals the end of the cooking time. The carry (really the borrow when counting down) can provide this. So that's what I did.

But that signal uses the complemented outputs of the flip-flops, which don't change at the same time as the uncomplemented outputs. This produced a glitch in the timer's Zero output. The glitch showed up only under a few conditions, but it was there nevertheless. The result was early termination of the run.

How is this fixed? I had to go back to the basic Down-5 and Down-9 Counters and install another gate to detect all 0s. It's on the lower right in Figs. 8.3-5 and 8.3-6.

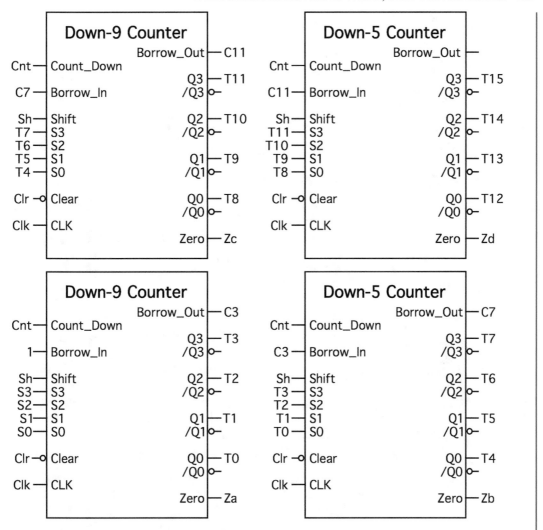

FIGURE 8.3-7: Timer.

That finished the timer and the clock. Now I have three parts finished. I decided I needed two more, one to handle the display of the clock and the timer and one to handle the keyboard and the buttons.

## 8.4 DISPLAY

The Display Driver is my circuit for displaying the timer and the clock; it's arranged to take either as input (Fig. 8.4-1). The choice of inputs is made via 2-to-1 muxes. The control signal is Disp_Tmr. When this is 1, the display is of the timer. Otherwise it is of the clock.

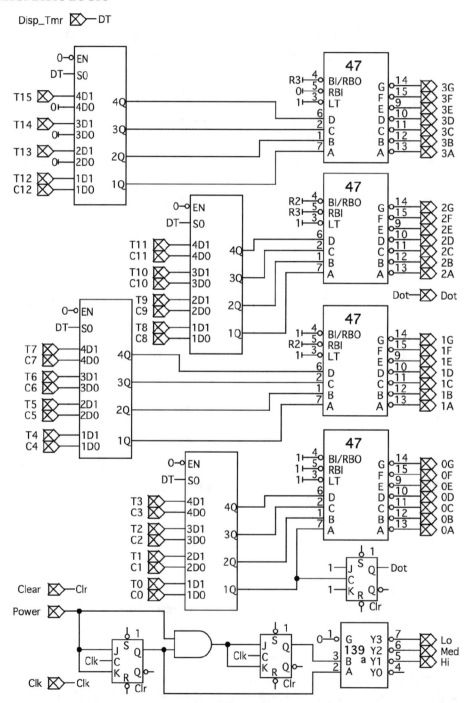

FIGURE 8.4-1: Display driver.

(The '47 part in Fig. 8.4-1 translates BCD digits into seven-segment displays. But it also includes "ripple blanking" so that leading zeros are suppressed. RBO, the "ripple blanking output," of the leftmost digit becomes the RBI, the "ripple blanking input," of the digit to its right and so on. The rightmost digit is not blanked. LT is the "lamp test" for turning on all segments.)

I also added a flasher (near the bottom right of the drawing) that would flash one of the dots (named DOT) in the display in step with changes in the rightmost digit. Why? Just to indicate that the thing is working, I guess. Most digital clocks do this. Looking back, I really should have made the dot flash once a second, which is not what it does.

This also seemed like a good place to put the circuit to select the power level. The two flip-flops at the bottom of Fig. 8.4-1 receive the signal from the Power button and provide signals for the three levels.

## 8.5   KEYBOARD CONTROL

The last module is for controlling the keyboard. I needed to do a number of things:

- Expand the keypad's strobe signal (the short line hanging down from the keypad symbol). This strobe is very short and not at the right place in the clock cycle.
- Convert the Clock-set, Stop-clear, Power, and Cook buttons to synchronized pulses. (I renamed Start as Cook.)
- Provide "seconds" and "minutes" pulses. so the clock and timer will run at the right speed.

The Keyboard Controller is shown in Fig. 8.5-1. Notice that I also used the Disp_Tmr signal to provide two different strobes from the keyboard (middle of the drawing). These will be used to control shifting through either the clock or the timer as one or the other is set.

I also decided to change the Zero signal from the timer into a pulse to avoid problems. If the timer reaches zero it will be stopped by removal of the Count signal. So Zero will remain asserted. Also, when the timer is cleared just before it is set, Zero will be asserted. I didn't want this signal held.

The combination of the '90 and the '92 divide the incoming pulse train by 60, producing a pulse once every 60 input pulses. The '90 is wired as ÷5 and the '92 is wired as ÷12.

Did the Keyboard Controller work? Yes, and so I converted in into a module. But as we'll see later, it wasn't designed quite right.

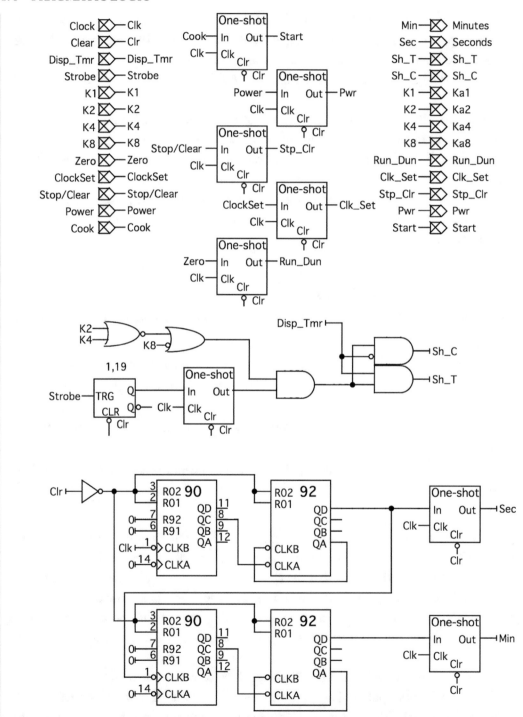

FIGURE 8.5-1: Keyboard controller.

## 8.6    COMPLETE SYSTEM

Now it's time to build up the complete system from the parts I've been developing. These parts are shown in Fig. 8.6-1. And since I have designed the parts carefully and tested them completely, the complete system should work correctly, right? It's OK to dream, if you accept the possibility that things just might not be perfect!

I started putting the whole thing together by listing the signals that I needed to provide. This list comes from looking at the inputs to the parts in Fig. 8.6-1. The table of Fig. 8.6-2 shows all the inputs to the five parts and where they come from.

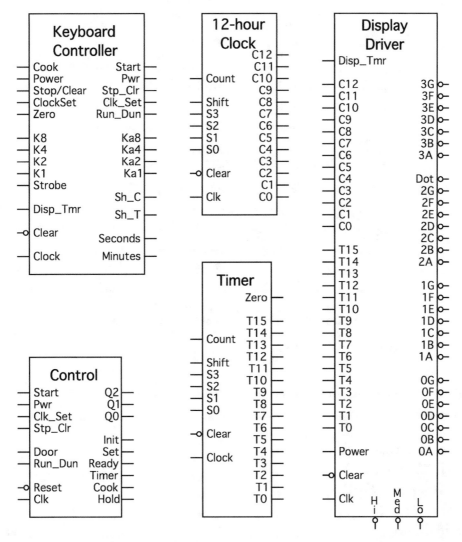

FIGURE 8.6-1: Microwave parts.

| Part | Input | From | Signal |
|------|-------|------|--------|
| Keyboard | Cook | Button | – |
| Controller | Power | Button | – |
| | Stop/Clear | Button | – |
| | Clock Set | Button | – |
| | Zero | Timer | Zero |
| | K8..1 | Keypad | – |
| | Strobe | Keypad | – |
| | Disp-Tmr | * | States |
| | Clear | Simulate | Switch |
| | Clock | System clock | |
| Control | Start | Keybd Contr | Start |
| | Pwr | Keybd Contr | Pwr |
| | Clk_Set | Keybd Contr | Clk_Set |
| | Stp_Clr | Keybd Contr | Stp_Clr |
| | Door | Switch | – |
| | Run_Dun | Keybd Contr | Run_Dun |
| | Reset | Keybd Contr | Switch |
| | Clk | System clock | |
| 12-hr Clock | Count | * | States |
| | Shift | * | States |
| | S3..0 | Keybd Contr | Ka8..1 |
| | Clear | * | States |
| | Clk | System clock | |
| Timer | Count | * | States |
| | Shift | * | States |
| | S3..0 | Keybd Contr | Ka8..1 |
| | Clear | * | States |
| | Clock | System clock | |
| Display | Disp_Tmr | * | States |
| Driver | C12..0 | 12-hr Clock | C12..0 |
| | T15..0 | Timer | T15..0 |
| | Power | * | States |
| | Clear | * | States |
| | Clk | System clock | |

\* Signal is derived from states from Control

FIGURE 8.6-2:  Signal inputs and sources.

There are nine signals in the table that have to be derived from the state of the FSM and other conditions:

Count (12-hour clock)
Shift (12-hour clock)
Clear (12-hour clock)
Count (Timer)
Shift (Timer)
Clear (Timer)

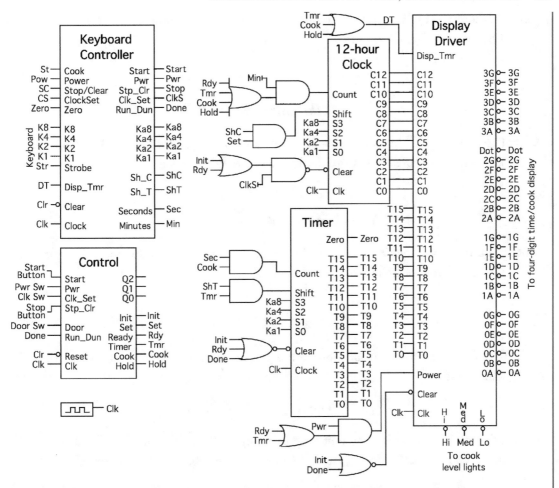

**FIGURE 8.6-3:** Finished microwave controller.

Disp_Tmr (Display Driver & Keyboard Control)
Power (Display Driver)
Clear (Display Driver)

I then thought about the conditions under which these signals have to be asserted. This gave each of the logic equations for those signals. Here are the results:

Count (12-hour clock): This clock is to run except in the Initial and Set Clock states. It is to tick forward once each minute. So it needs the Minutes signal from the Keyboard Controller. the logic is therefore

$$12hr\,Count = \overline{Init} \bullet (Ready + Timer + Cook + Hold) \bullet Min.$$

Shift (12-hour clock): The shift operation takes place on the strobe of the keypad when the FSM is in the Set Clock state. The Keyboard Controller provides the appropriate strobe, so the logic is

$$12hr\,Shift = Sh\_C \bullet Set.$$

Clear (12-hour clock): The clock to be cleared when the user presses the Clock Set button. But this is to be accepted only from the Initial or the Ready state:

$$12hr\,Clear = Clk\_Set \bullet (Init + Ready),$$

$$12hr\,Clock = System\,Clock.$$

Count (Timer): The timer runs only in the Cook state and requires the Seconds signal from the Keyboard Controller:

$$Timer\,Count = Cook \bullet Sec.$$

Shift (Timer): The keypad strobe shifts the digits in the timer when the FSM is in the Set Timer state:

$$Timer\,Shift = Sh\_T \bullet Timer.$$

Clear (Timer): The timer is cleared at the beginning by the Initial state, by Ready (since the next step is to set the timer), and by Run_Dun at the end of the cook period:

$$Timer\,Clear = Init + Ready + Run\_Dun,$$

$$Timer\,Clock = System\,Clock,$$

$$Timer\,Zero = Run\_Dun.$$

Disp_Tmr (Display Driver and Keyboard Controller): This signal determines which is displayed, the clock or the timer. The timer will be displayed during the Set Timer, Cook, and Hold states:

$$Disp\_Tmr = Timer + Cook + Hold.$$

Power (Display Driver): The power level is set when the Power button is pressed, but only in the Ready or Set Timer states:

$$Disp\_Power = Pwr \bullet (Ready + Timer).$$

Clear (Display Driver): This operation merely gets the part started and can simply be the Initial state:

$$Disp\_Clear = Init.$$

The final result is shown in Fig. 8.6-3. Rather than drawing lots of lines, I have used mostly names to link inputs to outputs. I added the Cook light to show that the microwave is operating. This would also be used to turn on the magnetron, the light in the oven, and the turntable.

And now the moment of truth. Did this circuit work correctly the first time? Why, of course . . . not. There were two problems:

- Clock and timer run in jerks—the problem was with the Minutes and the Seconds signals. These are derived by dividing the system clock by 60 and 60 again.

- But I forgot that "dividing" provides a signal that is 1 for half the period and 0 for half the period. In other words, the Seconds signal was 1 for 30 clock ticks and 0 for the next 30. Hence, the clock and the timer counted like mad for 30 ticks and then stopped for 30.

- Adding the one-shots to the Minutes and Seconds outputs of the dividers in the Keyboard Controller solved this problem. These provided output pulses only on the transitions of the dividers.

- Power level not reset at end of cook period—I simply forgot to reset the flip-flops in the Display Driver. Correcting this required a change in the logic for the Clear signal to add the Done signal to it:

$$Disp\_Clear = Init + Done.$$

Later on, when looking at what I had done, I realized that there is no reason to exclude the Initial state in the signal to make the 12-hour clock count. So I removed this unneeded signal from the Count logic for the clock:

$$12hrCount = (Ready + Timer + Cook + Hold) \bullet Min.$$

And now it *does* work.

## 8.7    SUMMARY

That's a pretty hefty design problem, especially for a beginning course. Can you do the same? I think so, particularly if you work in a team. Keep in mind that there will be a number of false starts and wrong turns. But try to reduce the problem to simpler modules that can be combined into the final project.

This controller has lots of individual chips. I didn't try to figure out how many, but I counted over 260 individual gates, flip-flops, and devices such as '47 drivers. That's lots of parts, even when they are combined into standard 74LS-series devices. Could there be a simpler way?

Yes, the microprocessor or microcontroller is the way this is done in all microwave ovens except those with mechanical timers. With just a few chips and a computer program we can build a microwave-oven controller. Moreover, we get more versatility and can add "features." Often the only thing that really distinguishes the top-of-the-line microwave from those lower down, other than cosmetics, is the features.

And that, said he, is that. Now it's your turn to do a substantial design. Good luck!

But wait! There's more! Do "real" designers design logic this way? Nope, it's all computers today. Now that you have the basics of logic design, go on to Chapter 9 for an introduction to Verilog-based design.

CHAPTER 9

# Large Systems: Do "Real" Designers Do Boolean?

How is it possible, using the techniques of the previous chapters, to design a large digital system? One answer might be, "Think really big!" But soon reality sets in: it isn't sensible to think about designing a million-gate microprocessor gate by gate. There has to be something better.

There is! Most design today is computer based. Languages that support this are called HDLs, hardware description languages. Two popular ones are VHDL and Verilog. Both were developed in the 1980s as larger and larger systems were being developed.

- VHDL stands for VHSIC HDL or very high speed integrated circuit HDL. (It seems strange to reflect speed in the name of a design language!) VHDL came out of Department of Defense efforts to develop a standard design language. VHDL looks somewhat like ADA and it has ADA's highly-typed structure. Because of this, it is considered harder to learn, but it is good for design of complex systems at the system level. IEEE Standard 1076 now defines VHDL.

- Verilog was proprietary until 1990, so designers were initially pushed toward VHDL. Verilog is now popular because it is easy to learn, partly because it looks like the language C. Verilog is not as well suited to handling complex designs at the system level. Its name apparently comes from *veri*fying *log*ic because it was originally developed as a simulation language. IEEE Standard 1364 now defines Verilog.

In this chapter, I'm going to present some Verilog code to introduce higher-level design techniques. My reasons for Verilog are in the previous paragraph: easy to use, looks like C. Be careful, though, because while it looks like C, it is *not* C and it is *not* a programming language.

*Not programming!* Let's emphasize that. Verilog is *not* a programming language. It describes hardware, hardware with gates and counters and wires and clocks. If you think of it as a programming language, you'll have trouble with some basic concepts such as continuous assignment.

What's continuous assignment? OK, when you write an assignment statement like A = B in any programming language, you mean to give to variable A the value that B has, but only when the sequence of steps in the execution of the program gets to this particular statement. Otherwise, the statement lies dormant.

In an HDL, when you write an assignment statement like A = B, you mean to have a wire connecting the source of the signal B to the destination A, continuously delivering the binary value of B to the device whose input is A. Continuously! Not just when this particular line of code is executed.

Why am I harping on this concept? Because you are more likely to run aground on this concept than anything else as you learn Verilog.

One more comment. This chapter is designed to just barely introduce you to modern high-level design through Verilog. When you want to learn more, see for example, Reese and Thornton, *Introduction to Logic Synthesis using Verilog HDL,* Synthesis Lectures on Digital Circuits and Systems, Morgan and Claypool, 2006.

## 9.1    BASIC VERILOG

Let's start into Verilog by looking at an example. I'm going to develop the same common, simple combinational-logic device using four different arrangements of Verilog code.

### 9.1.1    Structural Form

Figure 9.1-1 shows a Verilog module that defines my logic element. The module has two distinct parts, one defining what the module works with and one defining what work is to be done.

- The module statement defines the module by naming it and listing its input and output signals. Note that Verilog is case sensitive. Here, the key word is in lower case and its name is in upper case. While names can be in mixed case, I suggest not doing that: some synthesizers don't like it.

- The first input statement declares three one-bit signals that come from the outside world. Think of them as wires coming from somewhere else in a larger logic system.

- The second input statement declares a four-bit signal (in other words, four wires) with a common name. The individual signals are C[3], C[2], C[1], and C[0].

- The output statement declares a one-bit output signal, a wire going to somewhere in the larger system.

- In this example, some additional connections are needed within the module itself. These must be declared as well, but they are not connections that extend beyond the module itself. The wire statement declares four internal one-bit signal paths.

```
module MYSTERY1 (G, C, B, A, Y);

// Module for doing something

input G, B, A;
input [3:0] C;

output Y;

// Internal variables

wire P, Q, R, S;

// Logic in gate form

not (GB, G);
not (BB, B);
not (AB, A);
not (BBB, BB);
not (ABB, AB);

and (P, GB, AB, BB, C[0]);
and (Q, GB, ABB, BB, C[1]);
and (R, GB, AB, BBB, C[2]);
and (S, GB, ABB, BBB, C[3]);

or (Y, P, Q, R, S);

// What does this do?

endmodule
```

FIGURE 9.1-1:

- The remaining statements tell specifically the logic structure that is to process the signals. not, and, and or are three primitives in Verilog. They do just what their logic names imply. Each primitive's output signal is listed first, followed by the signals that are to be processed. For example, and (P, GB, AB, BB, C[0]); performs a logical *and* of the signals GB, AB, BB, and C[0], yielding the signal P.

- The five not primitives create the complements of five signals; the four and primitives each perform a logical *and* of four of these signals, yielding four intermediate signals; the or primitive performs a logical *or* of the four intermediate signals to produce the module's output signal.

- Finally, the endmodule statement is the other end of the wrapper around the module.

Neat, right? Very straightforward too. Easy to read. Two clear but separate parts, wrapped in a pair of statements. But what does this module do? Can you figure it out? Perhaps you could draw out the logic that the three groups of primitives define?

```
module MYSTERY2 (G, C, B, A, Y);

// Module to do something

input G, B, A;
input [3:0] C;

output Y;

// Logic in functional form

assign Y =  (~G & ~A & ~B & C[0]) |
            (~G &  A & ~B & C[1]) |
            (~G & ~A &  B & C[2]) |
            (~G &  A &  B & C[3]);

// What does this do?

endmodule
```

FIGURE 9.1-2:

Sorry! I'm not going to tell you yet what this does, because this code is not a good way to describe the operation that I want my Verilog code to instantiate (i.e., to create in hardware).

The particular form of Verilog code in Fig. 9.1-1 is called the "structural form" because it defines down to the last detail exactly how the gates are to be laid out and connected. While this is sometimes a good thing, most of the time it isn't, because we are not allowing the synthesis software the opportunity to do a better job of instantiating the logic. Perhaps the logic could be built more simply with fewer gates or shorter delays or . . . .

### 9.1.2   Dataflow Form

Figure 9.1-2 defines my same logic element but in a different manner. Instead of giving specific statements that tell exactly how to build the circuit, even to specifying the required parts, this example specifies the logic in a functional form that gives the synthesizer some latitude.

- The wrapper of module and endmodule is the same, with the same signals in the list.

- The signals coming and going are declared in exactly the same way via input and output. No internal signals are needed this time.

- It's in the description of the logic to be performed that this module differs considerably from the previous one. Here, the assign statement says to give the signal Y the logic value defined by the remainder of the statement. The tilde ($\sim$) means *not*, the ampersand (&) means *and*, and the vertical bar (|) means *or*.

This dataflow description tells what logic is to be instantiated without specifying exactly how to do it. Moreover, it's easier for us to read this code and figure out what is being done. Have you determined yet what common logic device this code describes?

### 9.1.3   Behavioral Form

The *structural* form of a Verilog description of logic hardware gets right down to the nitty-gritty of the hardware. It specifies exactly the composition of the hardware, giving the synthesis software no latitude whatsoever.

The *dataflow* form of a Verilog description offers some flexibility to the synthesis software. Rather than trying to tie down every detail of the implementation, it simply says, here is the output I want based on these conditions, using this general organization.

The *behavioral* form of a Verilog description of logic hardware says nothing about the organization of the hardware or even how the logic is to be designed. All it says is, here is what I want to have happen—you figure out how.

Figure 9.1-3 describes my same common logic device. This time, the code says to provide the output when certain things happen. For example, look for the always block in this code and notice that it says the output signal Y is to be connected to the input C[0] when the three-bit selection variable is all zeros.

Let's look through this code in more detail:

- The module—endmodule wrapper and the descriptions of the inputs and the output are the same as before.

- The output signal Y is given another attribute, reg. This means that the value (0 or 1) of the signal Y is to be retained at its present value if nothing comes along to change it. I'll say more about this in a moment.

- Wire creates an internal three-bit signal that is formed by the following three assign statements, yielding a single three-bit signal. This is merely to make it easier to write the code that follows.

- The always block is the workhorse of this code. It is saying, watch the signals sel and C. Any time either of them changes, update any signals that depend on them. In other words, continuously keep the signal Y up to date.

- The case code, with its wrapper endcase, tells what value to apply to the signal Y when the internal signal sel takes on a certain value. For example, when sel is the three-bit ("3") binary ("b") value 000, the output signal Y is to be connected to the zeroth bit of the four-bit input signal C.

```
module MYSTERY3 (G, C, B, A, Y);

// Module for doing something

input G, B, A;
input [3:0] C;

output Y;
reg Y;

// Internal signals

wire [2:0] sel;

assign sel[2] = G;
assign sel[1] = B;
assign sel[0] = A;

// Logic in behavioral form

always @ (sel or C)
    case (sel)
        3'b000:Y = C[0];
        3'b001:Y = C[1];
        3'b010:Y = C[2];
        3'b011:Y = C[3];
        default:Y = 1'b0;
    endcase

// What does this do?

endmodule
```

FIGURE 9.1-3:


- The last condition in case is default. This is needed to account for any conditions that are not otherwise included in the list. In this example, there are four possible conditions that could occur but are not specifically described, so default takes care of them. It is good practice to always include default even when everything has been specified because it prevents creating an unintended latch (an unclocked D flip-flop).

Do you know yet what this common logic device is?

No? Look at Fig. 9.1-4, where I have given the signals meaningful names:

- The signal G is now called ENABLE. It's an active-low signal, which you can figure out from the first part of the if statement in the always block: if it's 1, OUT is 0.

```
module MYSTERY4 (ENABLE, IN, SELECT, OUT);

// Module for doing something

input ENABLE;
input [1:0] SELECT;
input [3:0] IN

output OUT;
reg OUT;

parameter  IN0 = 2'b00, IN1 = 2'b01,
           IN2 = 2'b10, IN3 = 2'b11;

// Clearer behavioral description?

always @ (ENABLE or IN or SELECT)
    if (ENABLE == 1'b1) OUT = 1'b0;
    else case (SELECT)
            IN0 :    OUT = IN[0];
            IN1 :    OUT = IN[1];
            IN2 :    OUT = IN[2];
            IN3 :    OUT = IN[3];
            default: OUT = 1'b0;
        endcase

// What does this do?

endmodule
```

FIGURE 9.1-4:

- The two bits B and A, which specify which input is to be connected to the output, are combined into one two-bit signal called SELECT. This pair will be used with case.

- The four-bit input signal is called IN; the one-bit output signal is called OUT.

In addition to these changes, the four possible values of the SELECT signal are given names by the parameter statement. Note that these are merely names, not new signals. In other words, the two-bit combination 00 is now named IN0, and so on.

The always block is similar to the previous one, but the ENABLE logic is separated by an if–else pair. If the device is not enabled (ENABLE is high), the output is 0.

Have you figured out now that the device being described is a four-to-one multiplexer with an active-low enable? I hope you'll write Verilog code more like the last example than those before it!

## 9.2    FLEXIBLE VERILOG

What have we seen of Verilog so far? Actually, quite a bit of the stuff that relates to combinational logic:

- module and endmodule, the wrapper for a section of code;
- input and output to declare external signals, and wire to declare internal ones;
- not, and, and or gates;
- always @, which introduces code to provide continuous assignment of a value or signal to a signal;
- reg, which says that the signal's value is to be remembered as it is processed by the code, but not necessarily implying a flip-flop;
- assign to continuously assign the value of one signal to another;
- case and endcase, which provide multiple selections; and
- if–else, the basic conditional statement.

Now let's look at more combinational code in Verilog so we can see some other things it can do. More important, let's see how describing logic in Verilog allows us flexibility in constructing the code while allowing the synthesis software flexibility in constructing the hardware.

### 9.2.1    Coin Adder

The design example in Section 5.6 needed an adder that adds up the values of coins. I'll do a simpler version first, just adding dimes, nickels, and pennies to a four-bit register. (The largest sum is therefore 15.)

Figure 9.2-1 shows the Verilog module for this adder. The input signals are the current value of the four-bit count and three signals indicating which coin detector is asserted. This code presumes that only one coin detector is asserted at a time. The output of this module is the new value of the four-bit count.

The only new thing in this code is the use of + for addition. The nested if–else statement says to add to the count either 10 or 5 or 1. The code 4'd10 says that this constant is to be represented in four bits and that it has the decimal value 10.

Note especially the plus sign. We don't have to design the adder, we let the synthesis software do it. We have completely divorced ourselves from the details of how to do addition. The synthesizer is welcome to do straight addition with a ripple carry or to do carry look ahead. We don't have to worry about how it's done.

```
module COIN4 (CUR_CNT, D, N, P, NEW_CNT);

// Count dimes, nickels, and pennies

input [3:0] CUR_CNT;
input D, N, P;

output [3:0] NEW_CNT;
reg [3:0] NEW_CNT;

// Block adds 10, 5, or 1

always @ (CUR_CNT or D or N or P);
    if (D) NEW_CNT = CUR_CNT + 4'd10;
    else if (N) NEW_CNT = CUR_CNT + 4'd05;
    else if (P) NEW_CNT = CUR_CNT + 4'd01;
    else NEW_CNT = CUR_CNT;

endmodule
```

FIGURE 9.2-1:

This code uses an always @ block, so the output signal must be declared reg. This block continually "looks at" the signals in its list and continually provides the output value of the count through nested if–else statements. The last else statement tells what to do when no coin detector is asserted.

I used nested if–else statements instead of case, mainly to illustrate if–else. Using if–else can have implications during synthesis: if–else may be instantiated as a priority circuit, even if you didn't intend to assign priorities to the decisions. Using case avoids any such priority circuit.

Figure 9.2-2 illustrates the flexibility of Verilog. This new code implements the adder called for in Section 5.6. It's just a tweaking of the previous adder to use an eight-bit count and handle quarters, dimes, and nickels. There's nothing else new. Consider the effort you'd need to convert the logic diagram of Fig. 5.6-1 from eight bits back to four and you'll see the value of using a hardware description language.

## 9.2.2   Multiplication

A and B are two two-bit binary numbers; create C, their four-bit product. I'm going to do this in two different ways, one by essentially writing the truth table and one by manipulating the two-bit values themselves.

The first multiplier, Fig. 9.2-3, uses continuous assignment to develop the four bits of the product C. The four assign statements are just the logic derived from a truth table for the

```
module COIN8 (CUR_CNT, QUARTER, DIME, NICKEL, NEW_CNT);

// Count quarters, dimes, and nickels

input [7:0] CUR_CNT;
input QUARTER, DIME, NICKEL;

output [7:0] NEW_CNT;
reg [7:0] NEW_CNT;

// Block adds 25, 10, or 5

always @ (CUR_CNT or QUARTER or DIME or NICKEL);
    if (QUARTER) NEW_CNT = CUR_CNT +8'd25;
    else if (DIME) NEW_CNT = CUR_CNT + 8'd10;
    else if (NICKEL) NEW_CNT = CUR_CNT + 8'd05;
    else NEW_CNT = CUR_CNT;

endmodule
```

FIGURE 9.2-2:

```
module MULT1 (A, B, C);

// 2-bit by 2-bit multiply-- C = A x B

input [1:0] A, B;
output [3:0] C;

// Implement bit-by-bit logic

assign C[3] =  A[1] & A[0] & B[1] & B[0];
assign C[2] = (A[1] & ~A[0] & B[1]) | (A[1] & B[1] & ~B[0]);
assign C[1] = (A[1] & ~B[1] & B[0]) | (A[1] & ~A[0] & B[0]) |
              (A[0] & B[1] & ~B[0]) | (~A[1] & A[0] & B[1]);
assign C[0] = A[0] & B[0];

endmodule
```

FIGURE 9.2-3:

operation. (Verilog has provisions for actually typing in a truth table, but I won't do that in this chapter.)

The second multiplier, Fig. 9.2-4, uses a case statement to describe what to do with the input A under each of the possible conditions for the value of B. The braces { } are concatenation. For example, the second case says to concatenate two zeros and the two bits of A, which is the four-bit product of A with B = 01 in binary.

Because this second multiplier is in an always @ block, the product is declared reg.

```
module MULT2 (A, B, C);

// 2-bit by 2-bit multiply-- C = A x B

input [1:0] A, B;

output [3:0] C;
reg [3:0] C;

// Do arithmetic on A based on B values

always @ (A or B)
    case (B)
        2'b00 :  C = 4'b0000;
        2'b01 :  C = {2'b00, A};
        2'b10 :  C = {1'b0, A, 1'b0};
        2'b11 :  C = {2'b00, A} + {1'b0, A, 1'b0};
        default: C = 4'b0000;
    endcase

endmodule
```

FIGURE 9.2-4:

But why did I include default? It looks like all four possible selections for case are stated! The problem is that Verilog signals have four possible states, not just two. Two are obvious: 0 and 1. The other two are don't-care (x) and high-impedance (z). Since these aren't accounted for in the case selections, Verilog might object. Hence, it is generally good practice to include default, even if it appears that all cases are covered.

The codes MULT1 and MULT2 create the same logical operation, but the instantiated logic can be different. MULT1 specifies the form the logic is to take, specifying the details of the gating. MULT2 lets the synthesis software decide how to set up the logic. Both create the same logic, but one may be faster than the other or take less area on the chip. What actually happens depends on the synthesis software.

### 9.2.3   Priority Encoder

The last combinational-logic example is a priority encoder. This is a device that looks at the signals on several input lines and announces which signal line is asserted. If more than one line is asserted, the line with the highest priority is chosen.

The priority encoder in Fig. 9.2-5 has three input lines named, as a group, DATA. The line DATA[2] has the highest priority, so if DATA[2] is asserted, OUT is set to binary 10. If both DATA[1] and DATA[2] are asserted, [2] wins and OUT = 10 in binary. If no DATA line is asserted, the output NO is asserted to announce that fact.

```verilog
module ENCODE (DATA, OUT, NO);

// Three-bit priority encoder
// DATA[2] is high-order input
// Two-bit OUT tells which input is active
// NO says no input is active

input [2:0] DATA;

output [1:0] OUT;
reg [1:0] OUT
output NO;
reg NO;

always @ (DATA)
    begin
        NO = 1'b0;
        OUT = 2'b00;
        if (DATA[2]) OUT = 2'b10;
            else if (DATA[1]) OUT = 2'b01;
            else if (DATA[0]) OUT = 2'b00;
            else NO = 1'b1;
    end

endmodule
```

**FIGURE 9.2-5:**

Because this code uses an always @ block to create both OUT and NO, both of these must be declared reg.

The always @ block includes something new this time, a begin–end wrapper. The always @ block has several separate statements, so they need to be grouped to be inside the block. Begin and end do this. In previous examples, case and endcase form a single statement, as do cascaded if–else statements, so by themselves they don't need begin–end wrappers.

## 9.2.4    Assignment

A confusing feature of Verilog is blocking versus non-blocking assignments. The confusion comes when we forget that this is not sequential computer programming but instead is describing what is to become hardware. Let's look at these:

- Here's a *blocking* assignment:

$$B = A;$$
$$C = B;$$

- When this code is instantiated and becomes hardware, C will have the value of A. Note! A! The first statement connects signal B to source A. *Then* the next statement connects signal C to signal B. This is *blocking* in the sense that the first statement blocks the second until the first is instantiated.

- Don't forget that this is being done in hardware. Hence, the implementation of these assignments could be a wire-carrying signal A with taps along it leading to destinations B and C at the same time.

- Generally, good practice says that blocking assignments should be used in always @ blocks for combinational logic. All of the always @ blocks I have written so far are for combinational logic. How can you tell? The list that follows always @, which is the signals to watch, contains no clock signal. You'll see what that means in a moment.

- Here's a *non-blocking* assignment:

$$B <= A;$$
$$C <= B;$$

- When this code is instantiated and becomes hardware, C will have the *current* value of B, *not* the value after B is updated by A. This is non-blocking in the sense that both assignments are made at the same time, namely, the time of a clock tick. In hardware, this implies memory of some form such as a flip-flop.

- Generally, good practice says that non-blocking assignments should be used in always @ blocks for sequential logic, which explains the reference to the clock tick in the previous paragraph. We'll write always @ blocks for sequential logic in the next section. You can identify a sequential always @ block because it will have posedge or negedge in its list of signals to watch.

- *Blocking* assignments (=) should used in always @ blocks for combinational logic; *non-blocking* (<=), for sequential logic. They must never be mixed in the same block.

## 9.3 SEQUENTIAL LOGIC IN VERILOG

At the beginning of Chapter 6, I talked about the difference between combinational logic and sequential logic. I said that sequential logic remembers the past, but not too well. In that chapter and the next two, our sequential logic involved a clock; in fact, we called it clocked or synchronous sequential logic.

In this section, I am going to introduce some of the basic Verilog code for describing sequential logic. I'll stick to synchronous logic that will therefore always be characterized by a clock in the list in the always @ block. Activities specified in the block are triggered by one of

```
module COUNT1 (CLOCK, ENABLE, OUT);

// Three-bit clocked counter named OUT

input CLOCK, ENABLE;

output [2:0] OUT
reg [2:0] OUT;

always @ (posedge CLOCK)
    if (ENABLE) OUT <= OUT + 3'b001;

endmodule
```

FIGURE 9.3-1:


the edges of the clock pulse. I'll always choose the positive-going edge here, for no particular reason.


### 9.3.1   Simple Counter

Figure 9.3-1 shows the Verilog code for a very simple counter. This three-bit counter simply counts from 0 to 7 and then starts over. It has an active-high ENABLE signal—as long as ENABLE is high, the counter counts.

This code has three parts:

- Module–endmodule wrapper around the code. Notice CLOCK in the list of external signals. OUT is the counter's value.

- Input declaration of CLOCK and ENABLE and output declaration of the three-bit OUT signal, along with OUT's reg declaration. Any signal whose value is produced in an always @ block must be reg. In a sequential always @ block, this means the signal will be held in flip-flops. Hence when our counter is instantiated, it will probably have three D flip-flops in its hardware.

- Always @ block that now has a different-looking list of signals to watch. posedge says that any activity within the block is to take place on the rising edge of the CLOCK signal. Notice that no other signal is listed, even though OUT is used in the block. No signals need to be "watched" for activity because they are all checked on the rising edge of CLOCK.

How does this code work? The always @ block says it's to watch CLOCK. Every time a rising edge comes along, the signal OUT is to have 1 added to it, provided ENABLE is asserted. That's all there is to it!

```
module COUNT2 (CLOCK, ENABLE, CLEAR, OUT);

// 3-bit clocked counter with asynchronous CLEAR

input CLOCK, ENABLE, CLEAR;

output [2:0] OUT;
reg [2:0] OUT;

always @ (posedge CLOCK or negedge CLEAR)
    if (~CLEAR) OUT <= 3'b000;
    else if (ENABLE) OUT <= OUT + 3'b001;

endmodule
```
FIGURE 9.3-2:

You might notice that there is nothing that gives this counter a starting value. Right! There's not the usual clear or set or reset or anything like that. Is this a problem? Well, it depends on what you are trying to do. If you give this code to a simulator, the clock will never have any value; it'll remain undefined, at least with most simulators.

But remember that Verilog code is describing hardware. If this code is instantiated, it will be counter with real flip-flops. When the power is turned on, they will have some starting value. So the counter will count when asked to. That may not be what you want, but that's what will probably happen. However, you had fair warning when you tried to simulate the counter's operation. (Many synthesizers will warn you too.)

## 9.3.2   Better Counter

Let's improve on this counter by adding an active-low CLEAR signal to start the count at 0. Figure 9.3-2 shows the additions to the Verilog code to make this happen. One change is obvious: the addition of CLEAR to the input declaration. The other change is more complicated.

Notice that the always @ block now has a second signal in its list of signals. negedge shows that CLEAR is active low: it is to have its effect on the count OUT when it is goes low. This CLEAR signal is an asynchronous signal. Nope, I didn't say it wrong—it's asynchronous.

How can that be? First in the watch list of the always @ block is posedge CLOCK. This means that activity within the block is to happen on the rising edge of the clock. So far so good. Notice the next word in that list: or. This is saying that the next signal in the watch list is also to have an effect on activity in the block. Here, the falling edge of CLEAR is also to cause activity. This activity is triggered independently, not related to the clock edge. Hence it's asynchronous.

The always @ block is doing two things:

```
module COUNT3 (CLOCK, ENABLE, CLEAR, OUT);

// 3-bit clocked counter with synchronous CLEAR

input CLOCK, ENABLE, CLEAR;

output [2:0] OUT;
reg [2:0] OUT;

always @ (posedge CLOCK)
    if (~CLEAR) OUT <= 3'b000;
    else if (ENABLE) OUT <= OUT + 3'b001;

endmodule
```

FIGURE 9.3-3:

- When CLEAR falls, put zeros into the counter OUT (the if statement).
- Otherwise, when the CLOCK rises, add 1 to the count in OUT if ENABLE is high.

CLEAR wins in any race between it and CLOCK, so now we have a way to give our counter a starting value. This code will simulate properly and will produce properly working hardware.

Don't get confused here! Putting CLEAR in the watch list of the always @ block makes it asynchronous. To say this a different way, there are two signals in the watch list that control block activity and they do their jobs independently.

OK, so that brings up the question of how to create a synchronous CLEAR. The answer is surprisingly simple but sounds really strange: just omit CLEAR from the watch list! Figure 9.3-3 shows our same counter with a synchronous clear.

Huh? How can that be? Well, the watch list of the always @ block now is controlling activity in the block solely on the basis of the rising edge of the clock. This means all the activity is synchronized with the clock. Hence, the CLEAR signal's activity is synchronized with the clock. The statements in the block say, if at the rising edge of CLOCK the CLEAR signal is low, reset the counter. Otherwise, increase the count by 1 if ENABLE is high.

Just remember that asynchronous and synchronous seem to be intuitively reversed.

### 9.3.3    Another Counter

Here's one more counter, which implements in Verilog code, the counter of Section 7.1.1. It counts in binary from 1 to 6 and repeats. It has an active-high ENABLE and an active-low asynchronous CLEAR. I've written it with more "English" in the code to make the code more self-documenting.

```
module COUNT4 (CLOCK, ENABLE, CLEAR, OUT);

// State machine:1 - 2 - 3 - 4 - 5 - 6 - repeat

input CLOCK, ENABLE, CLEAR;

output [2:0] OUT;
reg [2:0] OUT;

parameter  ONE=3'd1, TWO=3'd2, THREE=3'd3,
           FOUR=3'd4, FIVE=3'd5, SIX=3'd6;

always @ (posedge CLOCK or negedge CLEAR)
    if (~CLEAR) OUT <= ONE;
    else if (ENABLE)
        case (OUT)
            ONE     : OUT <= TWO;
            TWO     : OUT <= THREE;
            THREE   : OUT <= FOUR;
            FOUR    : OUT <= FIVE;
            FIVE    : OUT <= SIX;
            SIX     : OUT <= ONE;
            default : OUT <= OUT;
        endcase

endmodule
```

FIGURE 9.3-4:

Figure 9.3-4 shows the code. I've given the six states names and then used a case statement to advance the counter. In the always @ block the first test is to see if the CLEAR signal is asserted and act on it if it is. The rest of the block advances the counter through its named states if ENABLE is asserted.

Don't forget that for synchronous logic, the watch list of the always @ block contains edge-sensitive signals. Activity in the block is triggered by a change of any of the signals in the watch list.

I have been calling the list of signals after always @ the "watch list." Technically, that's not the name given that list in the IEEE standard. It is properly called the "sensitivity list," or sometimes the "event control expression."

### 9.3.4  Moore and Mealy

Section 7.3 developed a state machine that can be instantiated either as a Moore machine or a Mealy machine. Do you remember the difference? Let's see if I do! A Moore machine has outputs that are associated to its states, while a Mealy machine has outputs that are associated

```
module STATE731 (CLK, X, CLR, OUT);

// Fig. 7.3-1 state machine (Moore)

input CLK, X, CLR;

output OUT;

reg [1:0] STATE;

parameter NO=2'b00, ONE=2'b01, TWO=2'b10;

// Moore output (state only)

assign OUT = STATE[1];

always @ (posedge CLK)
    if (~CLR) STATE <= NO;
    else case (STATE)
        NO      : if (X) STATE <= ONE;
                     else STATE <= NO;
        ONE     : if (X) STATE <= TWO;
                     else STATE <= NO;
        TWO     : STATE <= NO;
        default : STATE <= NO;
        endcase

endmodule
```

FIGURE 9.3-5:

to its transitions. In the state diagram of a Moore machine, outputs are in the state circles; in that for a Mealy machine, they are on the paths between states.

Figure 9.3-5 shows the Verilog code for the Moore version of this state machine. Its state diagram is Fig. 7.3-1. There's nothing unusual about this code—it has the usual wrapper, external signal declarations, and an always @ block with a clock.

I've declared (reg) a two-bit state variable STATE and then named the three states that are shown in the state diagram of Fig. 7.3-1. The always @ block uses a case statement to tell how to move from one state to the next depending on the input variable X. Notice that the active-low signal CLR, which resets the machine to the starting state (NO), is synchronous since it does not appear in the watch list.

One statement makes it evident that this is a Moore machine: assign OUT = STATE[1]. The output is associated to the state of the machine, so it's Moore. This assignment is outside the always @ block.

```
module STATE732 (CLK, X, CLR, OUT);

// Fig. 7.3-2 state machine (Mealy)

input CLK, X, CLR;

output OUT;

reg STATE;

parameter NO=1'b0, ONE=1'b1;

// Mealy output (state and input)

assign OUT = STATE & X;

always @ (posedge CLK)
    if (~CLR) STATE <= NO;
    else case (STATE)
        NO      : if (X) STATE <= ONE;
                  else STATE <= NO;
        ONE    : STATE <= NO;
        default : STATE <= NO;
        endcase

endmodule
```

FIGURE 9.3-6:

Figure 9.3-6 shows the same state machine done as a Mealy machine. This means the outputs are associated to the transition paths. Hence any output functions are controlled by both the state of the machine and the inputs. A check of the Mealy form of this machine in Fig. 7.3-2 shows that it can be done in just two states, which simplifies my code.

The assign OUT = STATE & X statement includes both the state of the machine and the input, so this is a Mealy machine. The state machine is in the always @ block, which follows the state diagram of Fig. 7.3-2. When CLR is asserted low, the next rising edge of the clock sets the state to NO. If CLR is not asserted, the rest of the block (else case) takes over and moves the machine between the two states.

## 9.3.5 Final Example

Section 6.4.3 presents the state machine for a railroad crossing signal. Its state diagram is Fig. 6.4-15. The Verilog code in Fig. 9.4-1 implements this machine. The state diagram is a Moore machine, so that's how I wrote the Verilog code.

```
module WVRR (CLK, RESET, N, M, S, FLASH);

// Whitewater Valley Railroad crossing signal
// N, M, and S are active low

input CLK, RESET, N, M, S;
output FLASH;

reg [1:0] STATE;

parameter NO=1'b00, INEND=2'b01, MID=2'b11, OUT=2'b10;

assign FLASH = STATE[0];

always @ (posedge CLK or negedge RESET)

    if (~RESET) STATE <= NO;

    else case (STATE)

        NO      : if (~N&M&S | N&M&~S) STATE <= INEND
                    else STATE <= NO;

        INEND : if (~N&~M&S | N&~M&~S) STATE <= MID;
                    else if (~N&M&S | N&M&~S) STATE <= INEND;
                    else STATE <= NO;

        MID     : if (~N&M&S | N&M&~S) STATE <= OUT;
                    else if (~N&~M&S | ~N&~M&~S | N&~M&S |
                        N&~M&~S) STATE <= MID;
                    else STATE <= NO;

        OUT   . : if (~N&M&S | N&M&~S) STATE <= OUT;
                    else STATE <= NO;

        default : STATE <= NO;

        endcase

    endmodule
```

FIGURE 9.4-1:

My code differs just slightly from Fig. 6.4-15—I had to use MID instead of M for a state name. The signals N, M, and S are all active low. Since the implementation of Fig. 6.4-17 uses flip-flops with asynchronous reset lines, I implemented the active-low RESET as an asynchronous reset. (Remember that placing the edge-sensitive RESET in the watch list of the always @ block makes RESET asynchronous.)

## 9.4   THE END

That's as far as I want to go with Verilog code. There are lots more things you can do with the system, and in fact you can get into all sorts of trouble! You have seen the basics and should be able to do both combinational logic and straightforward clocked sequential logic.

The Verilog language also has all sorts of constructs that cannot generally be synthesized to create hardware. One example is initial, which lets you write a block of code that initializes registers, etc. You can write the code, you can simulate it, but you generally can't reduce it to hardware through synthesis software. The code also includes such functions as multiplication; again, synthesizers often can't handle these. Be sure to look at the list of what your particular synthesis software can and can't handle.

Now it's up to you to go forward, if you wish, with digital logic design. If you want to follow the Verilog route, take a look at Reese and Thornton, *Introduction to Logic Synthesis using Verilog HDL,* Synthesis Lectures on Digital Circuits and Systems, Morgan and Claypool, 2006, a book in the same Synthesis Lectures series as this *Pragmatic Logic.* Try it! It's fun!

# Author Biography

Bill Eccles has been Professor of Electrical and Computer Engineering at Rose-Hulman Institute of Technology since 1990 (except for one year at Oklahoma State). He retired in 1990 as Distinguished Professor Emeritus after 25 years at the University of South Carolina. He founded the Department of Computer Science at that university, and served at one time or another as head of four different departments, Computer Science, Mathematics and Computer Science, and Electrical and Computer Engineering, all at South Carolina, and Electrical and Computer Engineering at Rose-Hulman. Most of his teaching has been in circuits and in microprocessor systems. He has published Microprocessor Systems: A 16-Bit Approach (Addison-Wersley, 1985) and numerous monographs on circuits, systems, microprocessor programming, and digital logic design. Bill has also published Pragmatic Circuits: DC and Time Domain, Pragmatic Circuits: Frequency Domain, and *Pragmatic Circuits: Signals and Filters*, three texts in this Synthesis Lectures in Digital Circuits and Systems series. Bill and his wife Trish have two children and three grandchildren. Bill is also a conductor (appropriate for an electrical engineer) on the Whitewater Valley Railroad, a tourist line in Connersville, Indiana. He is a Registered Professional Engineer and an amateur radio operator.